Water Auditing and Water Conservation

Water Auditing and Water Conservation

Jeffrey Sturman, Goen Ho and
Kuruvilla Mathew

IWA Publishing

Published by **IWA Publishing**
Alliance House
12 Caxton Street
London SW1H 0QS, UK
Telephone: +44 (0)20 7654 5500
Fax: +44 (0)20 7654 5555
Email: publications@iwap.co.uk
Web: www.iwapublishing.com

First Published 2004, Reprinted 2011

Printed by Lightning Source

British Library Cataloguing in Publication Data
A CIP catalogue record for this book is available from the British Library

Library of Congress Cataloging-in-Publication Data
A catalog record for this book is available from the Library of Congress

ISBN 978190222525
ISBN 1900222523

Contents

Preface

This book has arisen from a series of lectures in the unit 'Water conservation and water auditing', presented in the School of Environmental Science at Murdoch University, Perth, Western Australia. These classes arose from industry based intensive courses presented with the support of the Water and Rivers Commission of Western Australia to professionals in the mining, local government, horticulture and other large volume water consuming industry sectors, and the consulting professionals servicing these.

We found that while there was a considerable number of consultants and industry professionals actually conducting water audits, there was no systematic treatment of the subject for demand-side water auditing. Our lectures, and the unit materials or professional course materials of consultants and industry practitioners, have provided the background materials from which this book arose. While we have ourselves been the authors of the book, we have benefited a great deal from materials originally prepared by others based on their practical experience and expertise, and we acknowledge them below.

John Schlafrig, who was then with the Water and Rivers Commission, and John Brennan from the Water Corporation of Western Australia, were major lecturers at a professional course attended by two of us (JS and KM) in 1997. Their influence on us might particularly be detected in chapters 3, 4, and 5. We also made use of materials provided by John and Mel Rowe, of Rowcon Pty Ltd, in the preparation of Chapter 14, 18 and 19. Philip Hine of the Department of Environmental Protection prepared the materials for Chapter 9, Philip Commander of the Water and Rivers Commission materials for chapter 16, Kevin Lampard of Walkers Clean Water Company Pty Ltd for chapter 17, while Stuart Day, formerly with energy corporation Western Power, the first portion of chapter 18. Charles Buchnell of paper manufacturer Amcor Ltd provided materials for chapter 20. Marnie Leybourne of Water and Rivers Commission 22.2, while Alan Lindstrum, consultant to Water Corporation, 22.3. Teneal Davidson, one of our students, provided the report which became the first of the case studies in chapter 23.

We hope that this book will fill a gap in the professional literature and assist practicing auditors and cognate professionals as well as students taking our course and courses elsewhere. There is an Australian flavour to the book. This is appropriate we believe, as we live and work in the planet's driest significantly inhabited continent, where the practice of water auditing and the broad goal of water conservation are driven by necessity. It is in the area of the relationship of water auditing to water conservation and to environmental auditing that we believe this book starts developing a comprehensive framework in which to locate water auditing. Water auditing might be called 'quantitative water conservation'.

Jeffrey Sturman
Goen Ho
Kuruvilla Mathew

Murdoch University
Perth, Western Australia.
August 2004

The Authors

JEFFREY STURMAN

Jeffrey Sturman trained as a mechanical engineer at the University of Western Australia, taking the degrees BE and then PhD (1981) in the area of turbulent convective heat transfer. His engineering work has included air conditioning design, industrial noise and hearing conservation, lecturing in mechanical engineering, researching convective exchange flows (lakes and oceans) and more recently working in water auditing and conservation, with interests also in irrigation, hydrogen for energy, biomass for energy and other environmental technologies. Jeff is an ordained priest in the Anglican Church of Australia, and he has been involved in parishes, Anglican EcoCare and a range of educational and practical environmental projects. He is a Research Fellow at Murdoch University, where his current contact address is School of Environmental Science, Murdoch University, Perth, Western Australia, 6150.

GOEN HO

Goen Ho is Professor in Environmental Engineering at Murdoch University. He obtained his BE and MEngSc in Chemical Engineering from the University of Sydney and his PhD from the University of Queensland. His teaching and research interests cover a broad range of water and wastewater topics, and he has published extensively in these areas. More recently he has been attracted to the concept of sustainability. He is exploring the integration of water, wastewater, stormwater and water reuse systems together with the closing of water and nutrient cycles in engineered systems to achieve sustainability in the water sector. He also believes that it is important to link the water sector with the wider permaculture, renewable energy and sustainable building/ city sectors to achieve overall sustainability. He is active in the International Water Association, being chair of the Specialist Group on Small Water and Wastewater Systems and a member of the association's Strategic Council. Together with Kuruvilla Mathew and colleagues he has been instrumental in the establishment and development of the Environmental Technology Centre (http://www.etc.murdoch.edu.au).

KURUVILLA MATHEW

Kuruvilla Mathew is Director of the Environmental Technology Centre at Murdoch University. He obtained his BSc in Civil Engineering from Kerala University, DipSE from Delft International Courses in Sanitary Engineering, and PhD from Murdoch University. He has been passionate in his commitment to improving the water and sanitation conditions of developing communities including remote Aboriginal communities in Australia. He has developed his teaching and research interests and organised numerous local, national and international conferences in these areas. He is active in the International Water Association and is chair of the Specialist Group in Water and Waste Technology and Management Strategies for Developing Countries.

1

Introduction to water auditing and water conservation

1.1 WATER AND LIFE

Water is an essential ingredient for life upon earth. It comprises more than 80% of the mass of the human body and is necessary for the well being of approximately thirty million species inhabiting our planet Earth, with their accompanying ecosystems. Water appears to be abundant, with the oceans amounting to the order of 10^{18} m^3, a truly large quantity. Yet when we focus on human needs for water, this huge, readily available quantity of water is greatly reduced. Water can be used to satisfy just some human needs when it is saline to a substantial degree. For example sea water can be used for once through cooling in a conventional power station, yet mostly we need water which has a salinity of a few thousand parts per million or much lower than this. Drinking water, for example, should ideally have a salinity of less than 500 parts per million, though 1000ppm is often accepted, while 1500ppm is a convenient rule of thumb to identify brackish water. The availability of fresh water is truly restricted. Only about 2.7% of the world's surface water is fresh. Remove the ice component and the remaining fresh water amounts to only 0.8% of the total surface water. See Table 1.1. Our human life depends substantially on this relatively small quantity of fresh water. An encyclopaedic coverage of fresh water issues is available in the biennial publication on freshwater resources by Gleick (2000).

Table 1.1 Distribution of world's surface water by broad type

Type of water	Percentage of world's surface water of this type
Salt	97.3%
Glacial	1.9%
Fresh	0.8%
Total surface water	100%

The sheer number of human beings now alive limits the availability of fresh water to just modest quantities per person. Even these quantities are hard to obtain for many people. If we take 1700 m^3 per person per year to be the boundary below which water is regarded as being scarce, then water scarcity was experienced in twenty African and Middle Eastern countries in 1998. Further, 1.2 thousand million people in developing countries did not have access to safe drinking water (Cosgrove & Rijsberman, 1998). Safe drinking water is dependent upon biological purity as well as chemical purity. Water quality is an issue alongside water quantity. The number of human beings itself is a significant contributor to the biological, chemical and thermal degradation of water quality.

As a broad generalisation human beings are distributed on the Earth in a way that follows the spatial distribution of water. Where fresh water is readily available large numbers of people live. Thus human waste and industrial waste tend to be generated and disposed of near the source of water used by humans. While countries with advanced economies generally try to minimise the interaction between water and waste, countries with less advanced economies and burgeoning populations are often unable to protect water from the effects of the main consumers, or from discharges from the more limited numbers of (but massive) mining or manufacturing operations.

The ecosystems, which support life on our planet, all depend upon water whose quality is not significantly modified by human activities. Natural ecosystems have been substantially modified or even destroyed in areas of high human population concentration. The effects of humans are now being felt globally, even in areas of low human habitation or use.

1.2 A VISION FOR WATER AND THE WORLD

In order to structure further our discussion, we have selected a paper by Mahmoud A. Abu-Zeid (1998) who has identified an array of challenges facing the world in relation to water in a keynote address to the World Water Council in 1998. He characterised the world's fresh water situation as being in a state of crisis, exacerbated by climatic fluctuations. Throughout the next century he believes that current world water shortages will multiply quickly. While he identifies the challenges we have mentioned, water scarcity, lack of accessibility to clean drinking water and sanitation and deterioration of water quality, he identifies several others too, in company with many other authors (eg.Fullerton, 2001 and Anderson 1995).

Abu-Zeid (1998) believes water management is fragmented on a world-wide scale, with a vacuum at the apex of management and a confused array of weak institutions at the base. Financial support for a wide range of water supply, management, conservation and control matters is lacking to the extent of endangering public safety. The public, decision-makers, education systems, the media and others seem unaware of threats to the continued availability of good quality water, or that the sustainability of the planet is being threatened. Arising from these factors Abu-Zeid believes that there is a threat to world peace and security, based on an observation, among others, that political instability, tensions and public unrest characterise most countries facing water shortages.

In response to these challenges, the World Water Council (a peak body of water service providers) has determined a set of guiding principles to develop a vision for water in the world, for life and for the environment. Below we outline the guiding principles.

1.3 THE GUIDING PRINCIPLES

1.2.1 Assess fresh water resources

There remains a need to determine in an ongoing way the quantity and quality of fresh water and its distribution in time and space. Gaps exist in our current knowledge. Upon this knowledge we can base decisions to respond to the needs of both ecosystems and humans, within an overall commitment to sustainable development.

1.2.2 Apply integrated water resources management

An holistic approach to water resources management has been widely and well received in the past and remains an important component of the guiding principles. (Integrated water resources management means managing across water 'streams' and applications, accounting for the interactions between the various 'streams' and maximising system values not merely isolated values.)

1.2.3 Value water

Water may be valued from a number of different perspectives: cultural, social and economic (we add ecological too). Thus a unified worldwide valuation of water will need to be flexible to take account of the local physical, human and biological conditions.

1.2.4 Stakeholders in water share a common world interest

A foundation for such sharing of interest is a common awareness of the issues facing the world regarding water. Given the attainment of such common interest, the resolve of stakeholders to reach consensus will be greatly enhanced.

1.2.5 Stakeholders participate in international organisations

Participation (alongside a common interest) is important and equity in representation is necessary to foster worldwide ownership of outcomes.

1.2.6 Cover all forms of fresh water comprehensively and globally

Thus ground water and surface waters, liquid or solid phases, renewable and non-renewable sources and their inter-relationships are to be included in the vision.

1.2.7 Exercise will and commitment to translate vision to action

Will and commitment connect the vision to a set of targets, mechanisms and actions.

1.4 DISCUSSION ON GUIDING PRINCIPLES

We note that those who support ecologically sustainable development can readily embrace the guiding principles, and would perhaps add something to them. The principles focus particularly on fresh water. Many ecosystems and many human activities rely upon water that is not fresh. Even so the water requires certain characteristics for the ecosystem to remain healthy or for the human use to be efficacious. Seawater, which is important for both ecosystems and a variety of specifically human uses, deserves attention as well. So too do the variety of expressions of surface-water sources and ground-water sources, which though they are not fresh, are also very important for life.

The targets mentioned in the last of the guiding principles are identified by Abu-Zeid as follows: Clean drinking water and adequate sanitation, secured food supply, conservation of the environment and preservation of bio-diversity, sustainable economic growth and development and finally the promotion of world peace and security. Mechanisms are necessary in order to implement the vision with respect to the targets identified.

The mechanisms listed are certain to become incomplete as the world changes. Nevertheless specific mechanisms for the present are identified as: broadening the participation of stakeholders; raising public awareness; raising the capacity of institutions, staff and technological resources – especially within developing countries; development of technology through research and development – including conservation technologies and management systems and finally funding and mobilisation of finance.

The principles, targets and mechanisms above are directly related to water conservation and to water auditing. (For example the water management strategy, which arises as part of the standard water auditing process, is a small scale water management system.) The structured material we have presented provides background values and motivation for the subjects of this book. Particularly we locate our subjects within the mechanism 'development of technology', in our case intellectual technology. Yet this ought not to be taken in too restrictive a sense. As we will find later, water conservation and auditing engage the wider issues identified within the guiding principles, targets and mechanisms. For example, water auditing requires: enhanced awareness of the public and of decision makers, common and shared world interests, expanded capacity for dealing with water issues within and without enterprises (or even countries), to name just a few.

1.5 WATER CONSERVATION

Globally, water conservation means limiting or modifying the use of water by human beings, so that our use of water does not cause fluctuations of water quantity and quality within any cycle beyond those fluctuations caused by natural events (Imberger, J. 1997[1]) within the time-scale of human history. The definition applies to both global and local water cycles, to sources, reservoirs and sinks. This very general definition can be simplified in concrete situations and is often relaxed in its requirements. Water conservation in a local setting often means reducing the volume of water used for a single purpose or for complex systems of purposes. At the same time it implies improving the quality of discharge water and, if possible, reducing the quality of water required for the inputs to the purpose or purposes. By practising water conservation, we seek to provide social, economic, environmental and ecological benefits for the total global community of life, both now and into the indeterminate future.

Efficiency of water use is the complement of water conservation. In this case the quantity and quality of the water source is specified (likewise for the discharge or sink). The aim is to maximise the social, economic, environmental and ecological benefits from the given quantity and quality of water.

[1] We are indebted to Professor Jörg Imberger for the ideas behind this definition, which he explored in his search for a rigorous definition of sustainable development and which we have extended to water conservation.

Water management is concerned with water source investigation, water allocation and pricing and water source protection and, when protection has not succeeded, putting into practice remedial actions.

Practising water conservation, efficiency of water use and water management in general requires action by human agencies, communities and individual humans. These actions might include legislation or licensing, pricing and other economic instruments and human lifestyle and behavioural changes. We will look at an example.

We choose to focus initially on water resource allocation. Consider the situation where available surface water streams near a city are almost fully utilised by dams, where there is limited groundwater available and yet the city's water usage is rising. Legislation protects further surface streams but one from being tapped, with environmental and ecological protection being among the reasons that the legislation was enacted. Regulations strictly limit the amount of groundwater that might be withdrawn, to forestall the ingress of seawater into the fresh water aquifer. The water provider(s) to the city might increase the price of water in order to limit demand and to avoid the capital expenditure needed to dam the one remaining surface water source, which is allocated for water supply to the city. A consequence of this action will be a change in the behaviour of water users. People and businesses will need to change their lifestyles and modify habitual behaviour, which uses excessive water.

From this example we draw the conclusion that water source allocation implies a complex interaction between government, water supply providers, the human community, the environment and ecosystems. This complex interplay is a feature of all aspects of water management as well as both water conservation and water use efficiency. The example above, where economic instruments are used to compel a community to maximise the benefits from a given quantity of water demonstrate the interaction between water management and water use efficiency. They in turn interact with water conservation. Suppose the number of businesses grows in the city in the example; the businesses will then need to reduce their water use. While from the viewpoint of the city as an entity the issue is water use efficiency, from the viewpoint of the businesses the issue will become water conservation.

1.6 WATER AUDITING

Water auditing is the discipline concerned with quantifying water usage. It provides the means to develop precision in schemes for water conservation, water use efficiency and water management. Take some defined boundary around natural or human-made processes. (See Figure 1.1.) We assume that the system is at steady state and note that other balances are required for water quality assessment (silts for example).

The form of the quantitative heart of a water audit is to measure the quantity and quality of water outputs from within the boundary. The water outputs are compared with the water inputs; they should match within a pre-determined tolerance to regard the quantitative water audit as satisfactory. A water audit as we use the term is concerned with more than just this; water conservation and water use efficiency both inside and outside the boundary of the water audit domain are part of the water audit. In addition it will take account of water management issues outside the boundary as well as inside the boundary as appropriate. Viewing this another way, water auditing focuses upon the processes within the bounded region, with its inputs and outputs, yet it interacts with the wider world as appropriate. While water auditing is concerned with quantifying water usage, it is also value laden, relating to the vision, targets and mechanisms discussed above. Although water auditing was located in the 'mechanisms' above, it is related in a holistic way to the vision and to the targets.

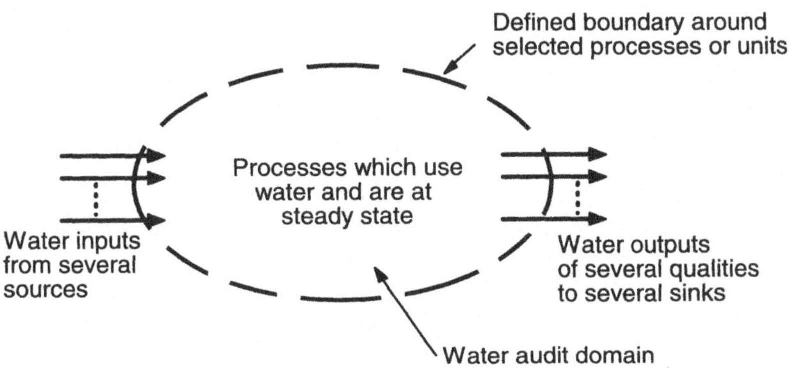

$$\text{Closure: } \frac{\text{The sum of water input quanities} - \text{The sum of water output quanities}}{\text{The sum of water input quanities}} < \text{predetermined tolerance}$$

Figure 1.1 The form of the quantitative heart of a water audit

The form of the quantitative heart of a water audit is to measure the quantity and quality of water inputs into the bounded region, evaluate the processes in which water is used within the boundary, and measure the quantity and the quality. Water auditing can be carried out in a wide variety of arenas. Natural systems such as a lake can be audited. The distribution systems of public water suppliers can be audited (This is sometimes called supply-side auditing). End users of water, whether domestic, commercial, industrial, mining or agricultural can be audited. Our focus within this book will be upon end users, with relatively little attention paid to the other categories. There is already work available to the public in the former categories. For example some issues of supply-side auditing are treated in a work by the American Water Works Association (1999).

Water auditing costs money; do the benefits of conducting a water audit outweigh the costs? In general terms we answer 'yes', though to quantify an answer requires each actual case to be examined. Why then do we answer 'yes'? In our experience and in the experience of colleagues with whom we have spoken, most enterprises, which use of the order of 10^5 kL of water per year or more, handsomely benefit from a water audit in purely financial terms and with a payback period on outlays normally of a few years at most. Smaller water users than this, including domestic users, will often benefit from a water audit, but the financial situation might require preliminary assessment before proceeding with a full audit. We answer 'Yes' for other reasons too, with benefits thoroughly canvassed in the outline on a vision for water and life. We have little doubt that the supply of water for human use will become more restricted in the future. The establishment of water auditing as a usual practice is highly desirable now, so that a bank of talent and experience is available for auditing when water supply becomes a yet more urgent problem in the future. Indeed water management is not possible at all without quantitative information. We believe that the same is true for water conservation and water use efficiency if they are to be taken seriously. Thus water auditing might be viewed as quantitative water conservation in the broadest sense.

1.7 REFERENCES

Abu-Zeid M. A. 1998 Water and Sustainable development: the vision for world water, life and the environment. *Water Policy*, **1**, 9 – 19.

Anderson, Terry L. (1995). "Water options for the blue planet", The True State of the Planet, Ronald Bailey (Ed), The Free Press, New York, pp.267-294.

American Water-Works Association (1999) *Water Audits and Leak Detection,* Hanover, 96pp.

Cosgrove, W.J. & Rijsberman, F.R. (1998) Creating a vision for water, life and the environment. *Water Policy*, **1**, 115 – 122.

Fullerton, Ticky (2001). Watershed, ABC Books, Sydney.

Gleick, P.H. 2000. *The World's Water 2000-2001 The Biennial Report on Freshwater Resources.* Island Press, Washington DC, 315pp.

Imberger, J. 1997 *Pers Comm.* The Centre for Water Research, University of Western Australia, Nedlands, Western Australia, 6907.

2

The relationship between water auditing and environmental auditing

In the previous chapter we examined water conservation and water auditing in the context of a worldwide vision for water and for life; we will now look at water auditing in a different context, that of environmental auditing. We expect initially that as concerns about water form a subset of concerns about the environment in general, water auditing will form a subset of environmental auditing. Environmental auditing is the subject matter of a series of environmental standards of the International Standards Organisation, the ISO 14000 series of standards. We have found that the relationships between the several ISO standards are easier to understand in visual form. Table 2.1 is a visual representation, which should be helpful in reading the remainder of this chapter. We include the contextual environmental management standards.

2.1 ENVIRONMENTAL AUDITING

At present there are just three standards specifically committed to environmental auditing. They are: ISO 14010:1996 'Guidelines for environmental auditing – General principles', ISO 14011:1996 'Guidelines for environmental auditing –

Audit procedures – Auditing of environmental management systems' and ISO 14012:1996 'Guidelines for environmental auditing – Qualification criteria for environmental auditors'. Of these standards, the first is general in nature, the second is specifically concerned with auditing pre-existing environmental management systems and the third, as its title indicates, deals with criteria for the appropriate qualifications of environmental auditors. All three are of interest to us in this book. Environmental auditing itself is defined in ISO 14010:1996 as a 'systematic, documented verification process of objectively obtaining and evaluating evidence to determine whether specified environmental activities, events, conditions, management systems, or information about these matters conform with audit criteria, and communicating the results of this process to the client'.

Initially, we note that the standards do not guarantee that an organisation's service or material product is environmentally acceptable, but rather they seek to provide a coherent framework for the on-going improvement of a system of environmental management (von Zharen, 1996). This has its counterpart in water auditing, where the water audit does not guarantee the environmental acceptability of an enterprise's service or material product. Rather the audit process seeks to evoke an ongoing improvement in the enterprise's conservation of water, water resource substitution, water use efficiency and quality of discharges to the environment and the financial performance of the enterprise. We notice too that the standards are not set within a prior grand vision for the environment, such as that outlined in the previous chapter for water. The standards are important instruments, with wide acceptance and they are of value in contributing to a systematic approach to water auditing. Our work will be informed by the existing ISO 14000 series of standards, which provide a framework for environmental management systems and environmental auditing alongside which water auditing can be set. Even so, we believe that specific water auditing standards should be added to the ISO 14000 series of standards.

Table 2.1 A summary of the pertinent ISO 14000 series of standards.

Standard	**Main subject**	**Main object**	**Specifies**	**Guides**
ISO 14001	Environmental Management Systems (EMS s)*		Yes	
ISO 14004	EMS s			Yes
ISO 14010	Environmental Auditing			Yes
ISO 14011	Environmental Auditing	EMS s		Yes
ISO 14012	Environmental Auditing	Qualifications for Environmental Auditors		Yes

* See text in § 2.2 below for a discussion on EMS s.

The only standard specifically (rather than generally) devoted to auditing, ISO 14011:1996, focuses on the auditing of an environmental management system. Water auditing too has a water management strategy associated with it and we discuss water management strategies in a later chapter in this book. We therefore find the standard ISO 14011:1996 interesting for our present discussion. ISO 14011:1996 relates closely with ISO 14004:1996, the standard on general guidelines for environmental management systems, and to ISO 14001:1996, the specification standard for environmental management systems. The relationship between environmental management system auditing and the associated environmental management system on the one hand, and water auditing and a water management strategy on the other hand is not identical. The word 'management' is used equivocally. In general the environmental management system audit takes place after the environmental management system has been set up, whereas the water management strategy in our usage of 'water auditing' is part of a water audit that has been undertaken initially. The water audit itself embraces most of the process that, in the ISO standards, is described as an environmental management system, which will be described below. Thus the use of the word 'audit' is also equivocal between the use in the ISO series and our experience of water auditing practice.

Of the other auditing standards, the specifically focused standard, ISO 14011:1996, has interesting detailed material for our purposes, though water auditing is not concerned with auditing the water management strategy. The generally focused standard ISO 14010:1996 has general material that we can apply to parallel processes in water auditing.

2.2 ENVIRONMENTAL MANAGEMENT SYSTEMS

In order to relate water auditing to the framework of the ISO 14000 series, we outline very briefly the environmental management system model of ISO 14004:1996 (an advisory document) and ISO 14001:1996 (a prescriptive document). The place of environmental auditing within the management system will follow. Five key principles summarise the management system. They are commitment & policy, planning, implementation, measurement & evaluation and review & improvement. (Ampersands link words composing one element.) We will tersely outline each in turn.

2.2.1 Commitment & policy

'An organisation should define its environmental policy and ensure commitment to its Environmental Management System.' (ISO 14004:1996) A documented review of the organisation's current status in terms of environmental protection is necessary prior to developing an environmental policy. Undertaking such a review presupposes a strong level of commitment from the organisation to improve the management of its activities, products or services from an environmental point of view. Especially the commitment of top management is essential and it is important that the commitment is widely characteristic of the enterprise.

2.2.2 Planning

A plan should be formulated by which an organisation's environmental policy can be fulfilled. The environmental policy will be the foundation upon which the plan will be based. The plan will take account of environmental aspects of the enterprise, which means activities, products or services that can interact with the environment. While there might be legal requirements with which to comply, it is desirable that internal standards be developed within the enterprise to assess performance of the enterprise in honouring its own policy. Overall aims, or objectives, are general, while targets are specific and able to be measured.

2.2.3 Implementation

Adequate capabilities and support mechanisms are necessary to implement an environmental policy, with its associated objectives and targets. Capabilities are dependent upon appropriately qualified and experienced people, availability of money and technical facilities and instrumentation. Implementation will be difficult if, within the enterprise, a set of environmental values at all staffing levels is not held in common with senior managers. Responsibilities need clear definition and paths of accountability need spelling out. Communication with internal parties is essential and with external interested parties is valuable in establishing an enterprise's environmental commitments. Legal issues might need consideration too. Operational processes are to be documented and procedures established for their control. By an operational process we mean processes or procedures which are defined clearly, properly documented and appropriately updated. The extraordinary needs to be considered alongside the normal, so that the enterprise can respond to emergencies.

2.2.4 Measurement and evaluation

An organisation's environmental performance requires measuring, monitoring and evaluating. An audit of the environmental management system is appropriate for this purpose, leading to preventative and corrective action to align actual performance with the enterprise's environmental objectives and targets and also with the law. An audit is carried out periodically by impartial, objective properly trained internal or external personnel. Thorough recording is necessary for both management systems and audits.

2.2.5 Review and improvement

With the purpose of improving its environmental performance, an enterprise needs to improve its environmental management system continually; this is a cornerstone of both ISO 14001:1996 and ISO 14004:1996. See Figure 2.1.

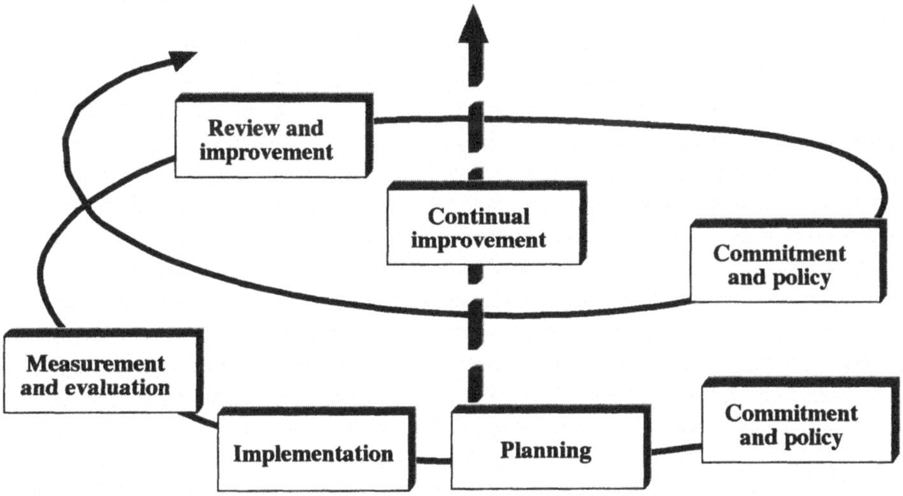

Figure 2.1 Environmental management system model – ISO 14001:1996 and ISO 14004:1996.

Using the meaning of 'audit' used in the ISO standards, we would locate a water audit as a component in the 'Measurement and evaluation' stage of an environmental management system. In the case of the first water audit of a series, the water audit might be a part of the planning stage for the whole environmental management system. Yet this is not the whole truth. As we noted above, a water audit usually includes the development of a water management strategy. We believe that the water audit is better viewed as a parallel part of the total environmental management system as indicated above. This implies that the definition of water auditing will need to go beyond the definition of auditing used in the ISO series. This will be picked up below. We reiterate that a water audit is not part of the audit of an environmental management system, but part of the environmental management system, which might be audited. The process illustrated in Figure 2.1 approximates the general process of a water audit as noted earlier. The water audit process is thus part of a water management system process, nested within and parallel to a similar process for environmental management systems on a larger scale.

To the above we add a word of caution. Water auditing is also concerned with the responsible use of water viewed as a global resource. Thus broader considerations are at stake than merely mitigating damage to the environment viewed solely locally.

2.3 AUDITING IN ISO 14010:1996 AND WATER AUDITING

Though the object of auditing in ISO 14011:1996 (environmental management systems) is different from our object (water auditing), the process of auditing outlined in ISO 14010:1996 and ISO 14011:1996 offers valuable material which we might assimilate to water auditing. In particular water auditing practice can benefit from the general standard, ISO 14010:1996. While deferring until following chapters the more detailed material we might choose to assimilate from ISO 14011:1996, here we will interpret the principles of ISO 14010:1996,

Prior to undertaking an audit the lead auditor, that is the person qualified both to undertake environmental audits and to manage them, should consult with the client (who commissions the audit). As a result of the consultation he should weigh whether information about the subject matter of the audit, water use in our case, is both complete and appropriate. The lead auditor should assess whether there are adequate resources for the audit to proceed and whether the auditee (the enterprise being audited) is likely to be sufficiently cooperative. The subject matter of the audit should be documented and clearly defined. Usually water auditors pursue necessary documentation when it is not available from either the client or the auditee. In any case the sources of the documents need to be identified clearly.

2.3.1 Scope and objectives

The boundaries and domain of the audit constitute its scope, which is determined by the lead-auditor, following consultation with the client, with the aim of fulfilling the audit objectives. The auditee should be made aware of both objectives and scope, in those cases where the client and the auditee are not the same.

2.3.2 Competence, independence and objectivity

In ISO 14010:1996 either internal or external audit-team members may be used subject to the client's desires. Note that an audit team might comprise one or more auditors (and could include those training to be auditors) as well as including technical experts. Together they should offer the skills, experience and knowledge appropriate for the particular audit. The audit will give rise to findings, or the outcome of evaluating data and evidence against the audit criteria. The auditor may draw conclusions from the findings when s/he applies reasoning to the findings. All parties require such conclusions to be objective, which assumes that audit team members are independent of what they audit. Conflicts of interest and conscious bias are to be avoided.

2.3.3 Professional care

Confidentiality is a key element of all professional practice including water auditing. No documentary or verbal issue from the water audit should be disclosed without the permission of both client and auditee, unless compelled by law. The conduct of the audit should be characterised by the skill, judgement, care and diligence that in similar circumstances would be expected of any auditor.

2.3.4 Systematic procedures

In general audits should be conducted according to systematic procedures, of which this book provides an acceptable model for water auditing. As we will see later, the detailed requirements of particular audits vary greatly, but the essential principles should remain.

2.3.5 Audit evidence, findings and criteria

The audit criteria (the requirements with which the auditor compares data and other evidence about the subject matter of the audit) are to be agreed by client, lead auditor, and auditee. We believe the auditee should be included unless legislation dictates otherwise. The quality of data and evidence, in our case mostly water flow quantities and qualities, should be sufficiently high that another auditor would make similar findings and draw similar conclusions were s/he to make an evaluation of the same data. Flow measurement and instrumentation will be covered in a later chapter. In the case of water auditing the criteria are not necessarily as clear as the standard implies and water auditing is a wider process as foreshadowed above and discussed below.

2.3.6 Reliability of audit findings and conclusions

Limited time and limited resources both imply that the data collected will be only a sample of that available. In water audits, the concept of closure, already illustrated in Figure 1.1, provides a quantitative assurance of accuracy. Yet closure itself has the limitation that the system for which input and output flows were taken is assumed to be at steady state. This assumption is not always true. Prominent disclosure is necessary of the tolerance chosen for closure with any implications. Issues presented by non-typical states of the audited system, together with non-steady state, must be disclosed.

2.3.7 The audit report

Though the audit report forms part of the section 'General principles', we will defer discussion of the topic until the later chapter devoted to the audit report.

We have now interpreted from ISO 14010:1996 material that applies to water audits. The summary above ought not to be seen as an acceptable substitute for reading the ISO standard itself. Our summary is merely intended to provide a framework for a partial understanding of water auditing.

2.4 THE GENERAL WATER AUDIT PROCESS

As introduced above, the definition of auditing in the ISO 14000 standards does not do justice to water auditing as a tool for water conservation and as a process parallel to the management process of Figure 2.1. We believe a description of water auditing is necessary that is more in accord with the actual practice and aspirations of water auditing.

Dawson (1997) has suggested a definition, having the basic structure of that of ISO 14010:1996, which we think is useful with further modification. With modification it is:

'Water auditing is a repetitive, systematic and documented process of objectively obtaining a balance between water input to, and water output from, an operation. Water quality is measured as needed. Opportunities are sought for a reduction of water use, for water reuse, recycling and for water resource substitution. Financial evaluations are made of all opportunities identified. A water management strategy is devised which is consistent with legal requirements, the enterprise's environmental policy and its movement towards sustainable development. The results of this process are communicated to the client and to the auditee where different'.

In Figure 2.2 we present a diagrammatic overview of the water audit process as practiced by the water auditor so as to flesh out diagrammatically the skeleton of the definition. Figure 2.2 is a reworked version of a similar figure in Dawson (1997). It is a simplified representation of the process as viewed by the auditor, rather than a global account. Phases one and two will be dealt with in detail in Chapter 3, phase three will be the subject of Chapter 4, and phase four will occupy Chapter 5 of this book.

Although the water management strategy that is adopted might become part of a wider environmental management system as discussed above, we think that the definition of water auditing uncovers an important fundamental difference between environmental auditing of the ISO 14000 series of standards and our approach to water auditing. The standard's definition of auditing investigates whether that which is being audited matches pre-established criteria. That is, it focuses upon deviation from criteria. The audit finds out what is wrong or what is right.

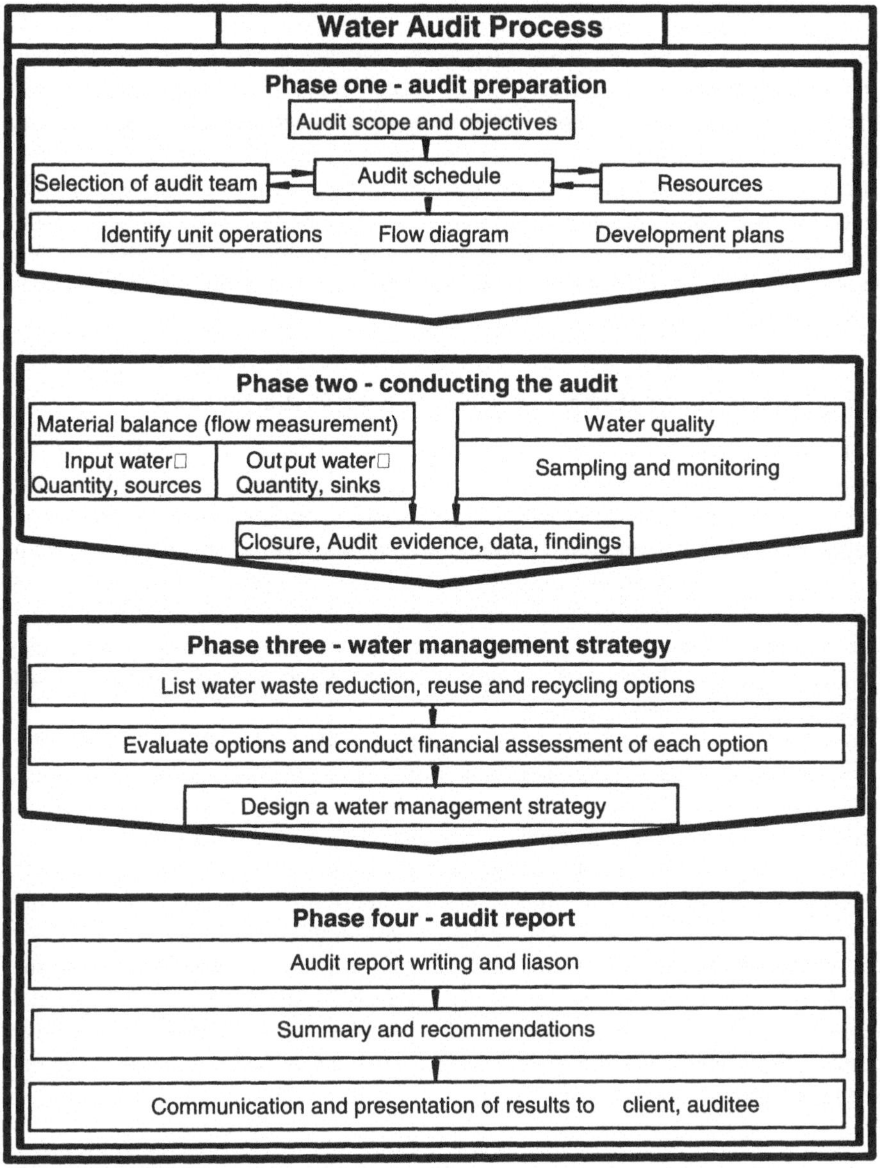

Figure 2.2 A diagrammatic overview of the water audit process as practiced by the water auditor. (This is a reworked version of a similar figure in Dawson 1997.)

Our approach to water auditing and that of our experienced academic and industry colleagues is to go beyond mere compliance with laws, regulations, external or internal standards in an enterprise. A water audit finds opportunities for improvement in water conservation strategies, water recycling and reuse, improvement of discharge quality and for improved financial performance of the enterprise subject to audit. Our use of 'auditing' includes that of ISO 14000 but goes beyond it. Thus our expectation at the start of this chapter that water auditing would be a subset of environmental auditing within the definition of ISO 14000 is not wholly sustained. This highlights even further the need for additional international standards specifically for water auditing.

2.5 QUALIFICATIONS OF THE WATER AUDITOR

ISO 14012:1996 is a guidance standard for criteria for qualification of environmental auditors. The standard lists five areas where their knowledge, skills and understanding should be addressed in work experience, formal training and maintenance of competence. We list the five areas in Table 2.2 in column one, with modified areas in column two to match more closely the needs of water auditors as we see them. Notice that we see the general requirements for water auditors as being very parallel to those of environmental auditors, with water auditors also needing familiarity with environmental science and technology, and with environmental management systems and standards.

Personal skills and attributes of the auditor are addressed in ISO 14012 too. We interpret them as good communication skills, personal and interpersonal skills, autonomy or the ability to be appropriately objective, personal organisation skills, deductive and inferential skills, and cultural and inter-cultural sensitivity.

We believe that these are essential for the water auditor too and should receive great prominence in planned experience, formal training and maintenance of competence. ISO 14012:1996 attends also to work experience and education, to training of auditors both on the job and formally. Most of the standard is very pertinent to water auditor training, yet some special requirements exist for water auditors. The standard requires that auditors have completed at least secondary education (or its equivalent). This requirement seems too loose for water auditors.

Table 2.2 Areas in which environmental auditors and water auditors should have work experience, formal training and should maintain their competence. Items in column one that have an asterisk are requirements for water auditors too.

Areas in which environmental auditors should have work experience, formal training and should maintain competence ISO 14012:1996	**Areas in which water auditors should have work experience, formal training and should maintain competence. (The items with an asterisk in column one are to be included too.)**
Environmental science and technology *	Water science and technology
Technical and environmental aspects of facility operations	Technical and water aspects of facility operations
Relevant requirements of environmental laws, regulations and related documents	Relevant requirements of water laws, regulations and related documents
Environmental management systems and standards against which audits may be conducted*	Water management strategies and professional and industry audit conventions in the light of which audits may be conducted
Audit procedures processes and techniques	Audit procedures processes and techniques

In our experience of teaching water auditing, students at second year university level often lack understanding of the principles behind facilities' operations and are at a great disadvantage in trying to learn technical aspects of facilities' operations. (See Table 2.2) We believe a tertiary degree is a necessary prerequisite for water auditors. Especially suitable degrees for potential water auditors are in mechanical, civil, environmental, chemical or agricultural engineering. Environmental Science courses also provide a suitable foundation. Physics or chemistry degrees likewise are suitable foundations. We teach water auditing in second year as part of an Environmental Science degree. Two years work experience with a water auditor is a satisfactory pre-requisite for certification as a water auditor by a professional body for graduates with such a degree.

Lead auditors are distinguished from auditors in ISO 14012:1996, requiring further experience. This seems entirely appropriate when applied to lead water auditors.

2.6 CONCLUSION

We expected that water auditing would be simply a sub-set of environmental auditing. This expectation was only partially met. Water auditing is a subset of environmental auditing in the sense that a water management strategy, arising from a water audit might become a component of an environmental management system. In so far as a water audit checks the quality of water discharges from an enterprise against criteria, it is part of an environmental audit. Water auditing goes beyond checking outputs from an enterprise against criteria. It is concerned with wider issues of water conservation, resource substitution, reuse and recycling and with achieving these alongside improving the financial performance of an enterprise. Thus we have developed a definition of water auditing which, while it contains elements of the ISO 14000 definition of auditing, embraces the wider concerns of water auditing as practiced. We repeat the definition here for the reader's convenience:

'Water auditing is a repetitive, systematic and documented process of objectively obtaining a balance between water input to, and water output from, an operation. Water quality is measured as needed. Opportunities are sought for a reduction of water use, for water reuse, recycling and for water resource substitution. Financial evaluations are made of all opportunities identified. A water management strategy is devised which is consistent with legal requirements, the enterprise's environmental policy and its movement towards sustainable development. The results of this process are communicated to the client and to the auditee, where different.'

We will revisit the relationship between environmental auditing and water auditing in Chapter 6, after the more detailed examination of the water auditing process in Chapters 3, 4 and 5.

2.7 REFERENCES

AS/NZ ISO 14001:1996 Environmental management systems – specification with guidance for use. Standards Australia and Standards New Zealand.

AS/NZ ISO 14004:1996 Environmental management systems – General guidelines on principles, systems and supporting techniques. Standards Australia and Standards New Zealand.

AS/NZS ISO 14010:1996 Guidelines for environmental auditing – General principles. Standards Australia and Standards New Zealand.

AS/NZS ISO 14011:1996 Guidelines for environmental auditing – Audit procedures – Auditing of environmental management systems. Standards Australia and Standards New Zealand.

AS/NZS ISO 14012:1996 Guidelines for environmental auditing – Qualification criteria for environmental auditors. Standards Australia and Standards New Zealand.

Dawson, M. 1997 *Water audit guide – Case study-Wesfarmers CSBP Limited Kwinana Works.* Independent study contract dissertation, School of Environmental Science, Murdoch University.

von Zharen, W.M. 1996 Understanding the Environmental Standards. Government Institutes Inc. 206p.

3

The water auditing process: audit preparation and conducting the audit

We introduced in Chapter 2 an outline of the water auditing process, which is summarised in Figure 2.2. Our intention in this chapter is to expand upon the parts of the water audit process labelled 'audit preparation' and 'conducting the audit' in Figure 2.2, using an amalgam of mainly our own material and some contribution from Dawson (1997). Phases three and four in Figure 2.2, the 'Water management strategy' and the 'Audit report' will be the subjects of Chapters 4 and 5 respectively. In Chapter 6 we revisit, extend and summarise the material of the last chapter.

3.1 AUDIT PREPARATION

3.1.2 Initiating the audit

A water audit might be undertaken in an enterprise for several reasons. An enterprise itself might initiate a water audit out of both its commitment to water conservation and its interest to reduce the enterprise's costs. Alternatively a regulator or Government department might initiate an audit to assist an enterprise to meet certain legal or regulatory requirements, or even to enforce compliance. Finally a water service provider might be interested to audit enterprises with large water usage, so as to reduce their use of water and thus defer the need for the provider to make large capital expenditures developing new water sources. These general reasons might be prompted by several more specific reasons:

- Desire for a green corporate image
- Community pressure
- Drought
- Increased demand outside enterprise might imply a higher unit price, and
- A desire to trade in water.

The initial phase of the water audit will be somewhat different depending upon which way the audit is initiated. In all cases the water auditor will expect a briefing document (the 'brief') to outline what is requested of the auditor. Table 3.1 illustrates the three possibilities that we have identified and who is consequently responsible for compiling the brief.

Frequently an enterprise will initiate a water audit without any sense of compulsion or policing by government agencies. Row one of table 3.1 shows the simplest distribution of varying roles, with the enterprise taking the roles of initiator, client, auditee and generator of the brief. In some instances the enterprise might not have internal staff qualified to write a brief for a water audit, so the enterprise might request the potential water auditor to write the brief on its behalf with careful consultation. We will discuss the content of the brief a little more fully below.

Table 3.1 Initiator of audits and the consequent identities of the client, auditee and who is responsible for compiling the brief.

Initiator of Audit	**Client**	**Auditee**	**Responsible for compiling Audit Brief**
Enterprise	Enterprise	Enterprise	Enterprise
Government agency	Government agency	Enterprise	Government agency
Water service provider	Water service provider and Enterprise	Enterprise	Water service provider and Enterprise

When a Government agency initiates the audit, we assume that there is some element of persuasion (or stronger) associated with the audit. An agency would be ill-advised to start its relationship with a forceful stance. They should do all in their power to assure the auditee that they seek to assist them to comply with either law or regulation. The outcome could well be to the auditee's financial benefit. Kudos might flow to the auditee following the satisfactory completion of the audit; the auditee might legitimately advertise itself as a responsible water conserving enterprise. In any event the agency becomes the client and the enterprise to be audited has a more passive role, though cooperation is still vital if an audit is to proceed satisfactorily. While the auditor's principal responsibility is to the client, the auditor needs to exercise all his/her interpersonal skills in relating to the auditee. If the auditor too can demonstrate that financial benefits will flow to the auditee from the auditing process, the auditor's task will be greatly simplified. This process ideally starts with the brief, initiated by the client, being discussed sympathetically and fully with the auditee.

As an initiator of an audit, a water service provider might be inclined to bring a persuasive stance. The water service provider might be in no position to provide even current levels of water supply unless either the provider expends large capital amounts, or end users reduce their consumption. A fruitful approach is to explore the common ground between provider and end user. Both might share or come to share common values to conserve water. Both desire to reduce their costs. Rather than the water service provider being merely the client, the provider and the end user could work as joint clients if an alliance of interests can be forged, with the end user enterprise also being the auditee. The development of the brief can be a joint effort, with financial responsibility for the brief remaining with the initiator, the water service provider. Such arrangements can greatly assist the relationship with the auditee.

The brief itself is a document that is developed to identify the scope of the water audit and the objectives. Typically the brief might contain requests to the auditor:

- to identify water usage from the sources used on the defined site, conveniently termed the domain, of the proposed audit,
- to attain a required accuracy for closure (ie. the pre-determined tolerance of Figure 1.1),
- to identify options for improving water use efficiency,
- to estimate the costs of these options,
- to categorise the options into required categories of payback period,
- to reduce water consumption individually or severally from all sources by greater than a nominated percentage,
- to determine whether the enterprise is complying with legislative, regulatory or license conditions for usage or discharge of water,
- to evaluate whether water management strategies from earlier audits have been applied and what the outcomes have been,
- to consider growth of water usage in the future, say over 10 years,
- to discuss all of the above with the client, before preparing the final audit report.

Definitions of terms might also appear in the brief.

3.2 RESOURCES

In response to the brief, the water auditor assesses the resources necessary to undertake the audit. The resources include money, people, time, instrumentation and data processing equipment. Clearly a visit to the site and discussion with the client and the auditee will be necessary at this stage. Some audits are sufficiently large to require a team to undertake the audit. Typically within such a team there will be a lead auditor, with subordinate auditors, students or trainees, as well as measurement, instrumentation or software experts. In some instances staff of the auditee enterprise might be part of the team. Where this is so there needs to be great care taken to ensure that no conflicts of interest occur. Not withstanding the above, an audit team of just one or two can undertake a surprisingly high proportion of audits. The audit is ideally scheduled on a time-line, which is mutually agreeable to the auditor, client and auditee. The fees charged to the client by the water auditor are to be agreed at this stage. The client's capacity to pay is one of the resources to be evaluated. A normal indicative charge would be 10% of the savings to the auditee resulting from an initial audit. This figure is somewhat arbitrary, but in practice often reflects a reasonable value for the work performed. A set fee would normally be charged for small, unusual or subsequent audits. At this stage the client and the auditor would normally enter into a formal contract.

3.3 WATER FLOW DIAGRAM

The water flow diagram, sometimes called a water management diagram, is the crucial systematising tool for the water auditor. We present in Figure 3.2 a sample water flow diagram for a typical shopping centre. Presently we are not interested in the details of a shopping centre except as they might help the reader understand water flow diagrams in general. The water flow diagram enables very physically complex systems of processes and pipes to be comprehended and inter-related in a way that assists the water auditor's task.

The water flow diagram shown in Figure 3.2 has a generally accepted form. On the left-hand side of the diagram the sources of water are identified, ground water and scheme water in this case. In the centre 'unit operations' (see Appendix 3.1 for detailed discussion) are represented as boxes, where a unit operation briefly means a generic use of water. For example, all showers are lumped together as one unit in our example. On the right hand side of the diagram the sinks for discharged water are identified – ground, of which some might be evaporated and some percolate to groundwater, sewer and storm water drains. Arrows on the lines represent the direction of water flow, while the quantities are identified in the units of thousands of litres per year (kL / yr). As several physical items with the same generic use are lumped together into one unit in the diagram, so each individual pipe is not shown, but one line on the diagram might represent several pipes with the same generic function. In principle the water flow diagram is simplified as much as possible without overlooking important units or lines of water flow, consistent with the scope of the audit. Water flow diagrams generally follow the layout in Figure 3.2, even for very much more complicated cases, where the central units of the process might take a much more complex arrangement than the neatly formed column at the centre of Figure 3.2.

Identification of general units (see appendix 3.1) or unit operations is an important first step in the process of gathering information towards constructing the water flow diagram. Such a unit is usually characterised by materials (including water) entering a domain, some operation taking place, and materials (including water) leaving the domain, normally in some manner different from the inputs. Thus in the Figure 3.2, staff showers have the inputs of clean water and dirty human bodies, the process of washing takes place and clean human bodies and dirty water are outputs. These building blocks of the water flow diagram are worthy of careful identification and comprehension by the auditor. Precisely what is an appropriate unit operation depends upon the scope of the audit. For example toilets and hand basins are not separated in Figure 3.2. In some circumstances they might need to be dealt with as separate units. Thus the blocks on the water auditing diagram might or might not correspond with a physical building like a toilet block or with a total manufacturing process, depending on the detail warranted by the scope of the audit as outlined in the brief.

A site visit is essential. The auditor collects information about general units or unit operations, and an array of other material, for the construction of the water flow diagram. Information might be sought about:

- site plans, including building layout and the plans of processes,
- drawings of pipeline layout,
- water sources, whether reused, recycled, seawater, surface water, ground water, storm water or scheme water,
- previous water audits,
- process water usage,
- general water usage,
- personal water usage points,
- reticulation,
- bore water usage,
- water treatment,
- water disposal, whether removed as product or discharged to ocean, sewer, ground water (recharge), watercourse or evaporated,
- plans for units or total processes to be developed in the future.

Interviews, questionnaires and personal inspection during the site visit are the processes used to construct the preliminary water flow diagram. The final water flow diagram requires determination of the flow rates and, where necessary, water qualities of the water streams. These measurements finally settle the importance of water streams and might allow the consolidation of some streams on the water flow diagram, while yet others might be ignored. There can thus be some degree of iteration between the notional audit preparation phase and the actual conduct of the audit to which we turn now.

3.4 CONDUCTING THE AUDIT

The essence of the 'audit' portion of the water audit process is to measure the mass (or volumetric) flows of all the flow lines in the water flow diagram. Additionally, the water quality should be measured in those instances where it is productive to do so. (More will be said about this below.) Upon completion of the measurements, closure is attempted, that is, (the sum of inputs minus the sum of outputs) divided by the sum of inputs is calculated as a percentage, see also Figure 1.1.

The standard accepted result is better than 10%. This percentage might be varied due either to requirements of the brief, or to experience during the course of the audit itself. We will expand on this below. Failure to attain closure warrants checking of calculations, further flow measurements, identification of pipelines as yet unknown, searching for leakages, or discovery of as yet unidentified cross linkages in pipes previously thought to be independent. The water flow diagram might be modified as a result of these actions.

3.4.1 Gathering the flow data – accuracy and tools

Physical metering and instrumentation is the first and obvious way of gathering water flow quantities. Chapter 7 is devoted to the detail of electronic instrumentation and flow measurement. Here we will deal mainly with broader issues.

3.4.1.1 Flow measurement and accuracy

The accuracy of the measurement appropriate for any flow-line is determined by what proportion the flow represents of the total flow through the audited domain. Thus a pipe representing 50% of the total flow might warrant accuracy of better than 5%, for example, whereas a line representing say 3% of the total flow through the domain could be measured with an accuracy of 50% with little consequence. This illustration ought not to be blindly applied. Actual circumstances could suggest a greater accuracy is required. For example, where there is an issue of water quality within a line, greater accuracy might be warranted than mere flow rates suggest. Generally the cost of purchasing or leasing measuring apparatus increases with increased accuracy, so there is financial incentive for selecting instrumentation of sufficient, but not excessive, accuracy. If earlier water audits have been undertaken they can provide indicators of flow stream magnitude, which can guide the current selection of flow instruments. Records could be available within the enterprise that perform the same function. A questionnaire (see below) might determine whether such records exist.

Many operations, which could form part of a water audit, will not operate at steady state. In these circumstances continuous electronic data logging could be appropriate. Integrating flow-meters might give satisfactory cumulative flows over a length of time at least as great as one cyclical period in reasonably consistent cyclical operations. Other situations might warrant the selection of sampling strategies informed by time-series analysis theory (see for example Bendat & Piersol 1986).

3.4.1.2 Questionnaires

Questionnaires might be distributed to those responsible for the unit operations or general units within the flow diagram, or to those with even wider responsibilities. The questionnaire could ask for initial estimates of flows through significant pipes, thus enabling intelligent selection of flow measuring devices for more accurate flow measurement. In some instances the estimate might prove sufficient without formal flow measurement. Questionnaires can also elicit information about pipe locations and significance, old flow records, inter-connections and water quality issues. The identities of key subordinates with special knowledge of the process or plumbing might be valuable too. It is most important to attempt to make such questionnaires brief and to the point, showing respect to staff for the heavy work-load they are likely to carry. Though the information derived from a questionnaire is most important, even more important is the cooperation a well-framed questionnaire might engender. There should be a short section explaining simply what the audit is about, with phrasing chosen not to alarm staff about their personal performance being under scrutiny. Indeed they should be affirmed as valued members of a team. These issues are best discussed with senior management and perhaps Union representatives prior to the compilation and distribution of the questionnaire.

3.4.1.3 Log sheets

When staff cooperation has been obtained, appropriate people might be willing to record flows at pre-determined times without disrupting work schedules. This process can result in considerable cost savings to the water auditor (and to the enterprise) as data logging computers and software costs are bypassed if this level of sophistication is not warranted. Where simple flow measurement means are adequate (eg. bucket and stop watch) selected staff could undertake the flow measurement as well as do the recording. Log sheets need careful and thoughtful preparation and explanation, so that the data required is what is actually acquired. Second requests are often less welcome than first requests.

3.4.2 Issues relating to input water from various sources

Input water can be water entering either the whole domain to be audited or entering an individual or several unit operations. In some circumstances the volume of water required can be lessened by input water of higher quality. There is some interplay between requirements of water quantity and quality and therefore both the quantity required and the quality of supply can impact upon the selection of water supply sources.

3.4.2.1 Ground water, surface water, sea water

In some parts of the world use of ground water and surface water will be controlled by either regulation or legislation. Part of the water audit could therefore be to determine whether the local regulatory or legislative requirements are being met. In any case the flow rate of the ground or surface water input to the domain of the audit is required. While it is possible to make flow measurements from the discharge side of a pump, there is a widely used short cut. Where conventional electric pumps are extracting ground water, surface water or sea water, and delivering it for its purpose, it is often sufficient to log the hours the pump is run. With a typical discharge pressure, assumed steady, the pump specification curves can reveal the pump output flow-rate. Usually this is quite accurate enough for the water audit. Interesting variations to this common arrangement can be found in the field. We will look at just one example, to illustrate that short cuts must be carefully evaluated before being applied blindly.

Increasingly in remote areas individual pumps are being operated by dedicated solar panels. If energy is stored in batteries and then delivered to an electric pump, the same process can be applied as described above. Sometimes solar panels are connected via a conditioner directly to a DC motor driving a pump. In this instance the above procedure certainly cannot be applied, as the pump does not operate at a fixed point on its performance curve. Frequently, a water storage tank accompanies such an arrangement. Water input to the audit domain can be determined by shutting off the input pump and measuring the decreased volume of water in the tank over an appropriate period.

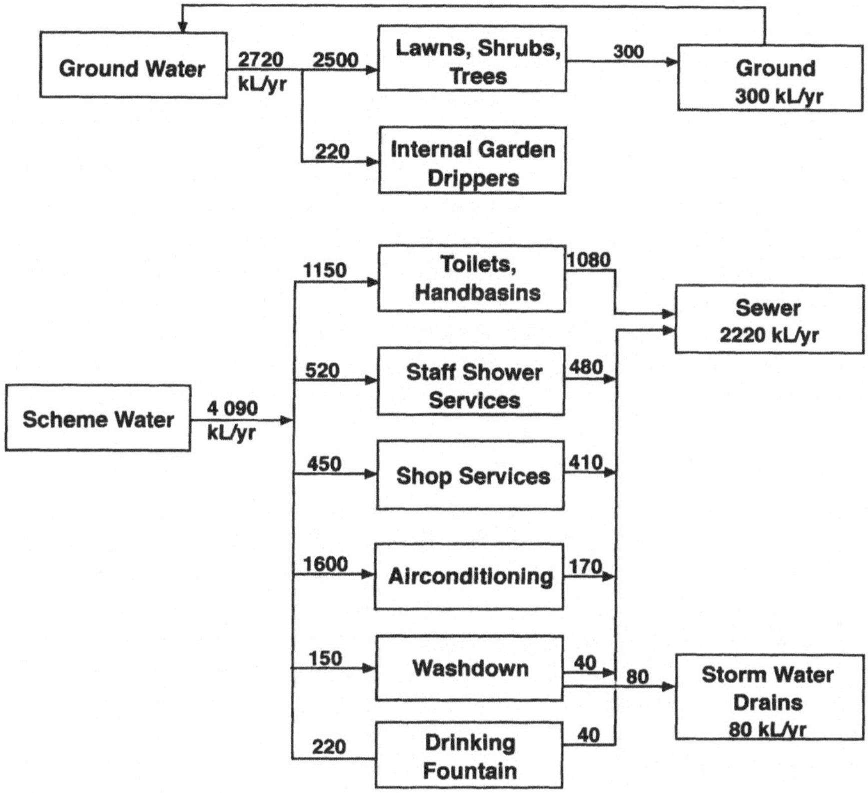

Figure 3.2 Typical shopping centre flow chart. Water sources are on the left, unit processes are in the centre and water sinks are on the right. Arrows show the direction of flow, and flow volumes are shown in the units of kL/yr. (*Adapted from a flow diagram of the then Water and Rivers Commission, Government of Western Australia.*)

3.4.2.2 Scheme water

Water service providers generally provide a meter at the (water) entry to serviced premises. This meter usually accumulates total volume flow. By choosing intervals in time at which the meter is read, the meter can be used to gain water flow rate information averaged over time-scales ranging from the length of the audit down to any sampling rate the auditor chooses. Gaining mean flow rates over small time intervals is more expensive than gaining them over longer intervals. The accuracy required will be determined by the scope in the brief and conditions found in the course of the audit.

An alternative method of gaining scheme water flow rates is to use information from water bills. Disadvantages of this method are that the period over which the flow rate is averaged depends upon the billing period, normally determined by the water service provider, and that the billing period probably doesn't correspond with the period of the audit.

3.4.2.3 Rainwater and stormwater

Rainwater and stormwater inflow to a site might represent an opportunity for water resource substitution. The potential rainwater available is easily determined by multiplying rainfall rates by the surface area of the site (assuming no gross slopes). On a smaller scale than a site, rainfall from a building's roof, or from a bituminised surface, could be calculated similarly. We would classify rainwater collected from a tarmac as stormwater. Run-off or stormwater from natural ground surfaces is complex to calculate; we recommend the services of a surface-water hydrologist if it is to be determined seriously.

3.4.2.4 Recycled water

In Figure 3.2 there is shown a feedback loop from water discharged to the ground and water input to the top two unit operations sourced from groundwater. In this case the feedback or recycle loop extends over the whole audit domain. In some instances, recycled water streams can occur within a unit operation or over more than one unit operation, while still remaining well within the extremities of the audit domain. It is a goal of a water audit to use such recycling where possible. The water auditor needs to exercise caution in conducting the audit not to attribute water from a recycled stream to normal input to the unit operation.

3.4.3 Issues relating to water discharged to various sinks

There are many discharge routes that might be pertinent for a single unit operation or for a whole audit domain. They include the domain's product (eg soft drink), waste product (which might be recovered), water which might be reused (possibly with treatment) in other unit operations, or recycled (possibly with treatment). Water might be loaded with heat and discharged to cooling ponds, or loaded with contaminants and be discharged to the sewer (often at great cost). Water can change phase and might leave the unit operation or the audit domain as steam or even as ice! In addition, discharges from the audit domain as a result of storms should be considered and more will be said about this in Chapter 4. Measuring the discharge from these various routes is often more of a challenge than measuring the input water flows.

3.4.3.1 Estimating discharges

We will illustrate this issue with a simple example, using as an example the miniature unit operation of a single shower. Clean water and a dirty human being enter the process; dirty water and a clean human being leave the process. We envisage that the discharge water is destined for a sewer. It is indeed possible to measure the flow from the floor drain into the sewer. Is it necessary within the accuracy required for the audit and does it warrant the expense? In many cases the answer will be 'no'. Water is lost due to evaporation, both directly while the shower is in progress and indirectly from a towel used to wipe a wet body after the shower. Splashes on the floor will likely evaporate too or aggregate to a size where they will run down the drain. These losses are well below a typical audit closure tolerance. Therefore the discharge to the sewer is often measured by measuring the flow rate from the shower head using a bucket and stop watch, assuming a mean shower length and obtaining the number of showers the head services in one day. *Thus the input flow rate to the unit operation is assumed to be the same as the output flow rate within acceptable limits.* Although this example appears trivial, there are important matters flowing from it.

Seldom would a single shower be designated a unit operation in a real-life audit. Normally there would be a large aggregation of showers and perhaps hand-basins and other processes into a general or unit operation. We recall too that more biologically active materials than shower waste discharge into a sewer. One of the purposes of the audit process is to detect leaks, which is one of the possible causes for failure to obtain closure. In the case of discharges into a sewer, the possible impact of leaks into the environment is not benign. When estimates such as those outlined above are used for the discharge to sewer, leaks will not be indicated from failure to obtain closure. Thus extending from our simple example, we draw the conclusion that not measuring actual discharges disables one instrument for indicating leaks. We also draw the reader's attention to the fact that while in principle the 'audit' portion of the water audit process (see Figure 1.1) compares discharges with inputs, within a specified tolerance, often water audit practitioners reduce the pain and the cost of the audit by taking short cuts. Intelligent estimates are made of discharges, yet not without risk. We believe such short cuts are generally justified, but they should always be taken consciously and identified within the auditors mind and also, when the risks are important, in the audit report (see Chapter 5).

3.4.4 Measuring water quality

Water quality is measured in a water audit only as necessary. Often, where its measurement is necessary, the enterprise whose premises are being audited will already have a measurement regime in place and could have its own water quality laboratory. We outline typical characteristics that might be measured. Nutrient concentrations and the related Biochemical Oxygen Demand (BOD) are important for water to be discharged into waterways, as also is the concentration of heavy metals. Both temperature and pH of discharged water also have implications for both physical and biological environments. Total Dissolved Solids likewise have ramifications for the environment, along with Total Suspended Solids. Part of the water audit process is to consider all water discharges as potential sources of water, with or without treatment. High chloride concentrations in water can cause problems for pressure vessels, especially those at high temperatures. Thus chloride concentration is another particular water quality parameter of interest to the water auditor. Legislative or regulatory limitations might be imposed on the quality of water discharged to sewers (e.g. oil or fat levels), surface waters and streams, to ground and to the ocean. Charges are sometimes applied. In such circumstances measurement of water quality is compelled by legal and financial force. Water quality will be discussed in Chapter 8.

3.4.5 Closure

Again we refer to Figure 1.1, which illustrates the basic idea behind the 'audit' aspect of water auditing. As we have foreshadowed almost immediately above, the field reality of water auditing procedures is not always as clear as the figure suggests. The basic idea is to use a comparison between water input to an audit domain and water output from the domain to alert us to leaks, overlooked water streams, unknown inter-connections within the domain or perhaps even to unknown pipelines extending beyond the audit domain. Closure, that is obtaining the comparison of water input flow rates and water output flow rates within a prescribed tolerance, is a tool to help us be sure we are accurately describing actual water usages and stream flows. The tool is not foolproof, as closure may be obtained satisfactorily when stream flow measurements are incorrect but errors are compensatory, that is, in opposite directions. In order to refine closure as a tool, we can apply it in more ways than the merely gross way illustrated in Figure 1.1. Closure can be sought for the whole audit domain (eg Figure 1.1), and it can be sought between water input to the domain and the points of water usage. Also we might seek to close between the points of water usage and discharge from the domain, e.g. in Figure 3.2, wastewater auditing would constitute an audit from unit operations to sinks. Wastewater auditing will be revisited in Chapter 21. Especially where water quality is an issue, closure can be sought for a unit operation. We determine where to seek closure by assessing the requirements of the brief and the conditions at the site.

3.5 REFERENCES

Bendat, J.S. & Piersol, G 1986 *Random data: analysis and measurement procedures.* New York, Wiley, 2nd edn., 566pp.

Dawson, M.D. (1997) *Water audit guide. Case study - Wesfarmers CSBP Limited Kwinana Works.* Independent study contract dissertation, School of Environmental Science, Murdoch University, Murdoch, Western Australia.

Shreve, R.N. & Brink, J.A. Jr. 1977 *Chemical process industries.* McGraw Hill.

APPENDIX 3.1 UNIT OPERATIONS

The term 'unit operation' has its origins in the chemical industries. A short history of the term will be found in Shreve & Brink (1977) for example. The essential idea is that in a complex chemical industry, there are common simple operations for acting upon a chemical stream, regardless of the chemical detail of the whole process. For example 'stirring' is necessary in many chemical processes. It is a unit operation. Usually unit operations involve physical change rather than chemical change, which was formerly called a process, though other terms might be used. Water auditors have picked up this terminology in a somewhat inexact way. Each of the boxes in the flow diagram we call a 'unit operation' in the sense that it represents a simple operation upon water and usually something else. Behind the detail of what a unit operation might mean lies the simple desire of the water auditor to simplify very complex systems of water use without throwing out any essential operation, attention to which might enable savings in water use, reuse, recycling or water resource substitution. In Figure 3.2 we illustrate the tendency for water auditors to lump together more than one operation for the sake of convenience; the 'unit operation' 'toilets and hand-basins' might arguably be two unit operations and require two boxes, but we use only one for it is of little consequence for the outcome of the audit and it conveniently summarizes operations closely connected, often in the same physical space. This box we call a 'general operation'.

4

The water auditing process: the water management strategy

We continue working through the simplified schematic outline of the water auditing process shown in Figure 2.2. The water audit process progresses from the satisfactory attainment of closure, or of a series of selected closures, described in Chapter 3. We should therefore have a sufficiently accurate water flow diagram, which will form the foundation upon which we build. The next procedure, which takes place in the water auditing process in the general sense, is the development of a water management strategy. Developing the water management strategy can be the most creative portion of the total process, in which our general aspirations for water conservation are addressed in the local domain of the water audit. There are conceptual tools, which we can use to seek ways of reducing water use, identifying water reuse and recycling options and attaining waste discharge reductions. We will start our discussion by outlining the meaning of a set of conceptual tools and then move to a more focused discussion of how the tools might be utilised. Introductory notes by Schlafrig (2000) and an expanded form in Dawson (1997) have informed our thinking.

4.1 CONCEPTUAL TOOLS

4.1.1 Environmental management systems, Cleaner production

We have discussed environmental management systems briefly in Chapter 2. Such systems (eg. ISO 14004:1996) provide a broader environmental context in which a water audit might be conducted. In turn, the development of an environmental management system could itself be informed by other conceptual tools such as Cleaner Production. Cleaner Production is (see United Nations Environment Programme, 1996) 'the continuous application of an integrated preventative environmental strategy to processes and products so as to reduce the risks to humans and the environment'. It is very important for a water auditor to be aware of these tools, as too singular a focus on merely conducting a water audit could actually lead to recommendations detrimental to the environment when viewed more broadly. We recommend that the water auditor be sufficiently familiar with these concepts to be able to discuss with senior management of the client and the auditee a prospective water audit within their context at the time of receipt of the audit brief.

4.1.2 Zero discharge

Unfortunately there is not general agreement about the meaning of this concept. In the European Community, for example, the generally accepted meaning is ' no continuous discharge during dry weather, allowing intermittent discharges, but rainfall run-off is allowed to occur '. By contrast, the accepted meaning in the United States of America and also in Australia is ' no discharge at all during dry weather either continuous or intermittent until after a storm of prescribed intensity and duration '. The intensity of a storm is usually expressed in mm of rain per hour and the duration in minutes or hours. The latter definition is more rigorous than the former. The concept ' zero discharge ', despite the title, does not imply that no water leaves the site. For example water might leave the site by evapo-transpiration or by evaporation. Clean water can leave the boundaries of a licensed site.

The water auditor's task includes being familiar with the appropriate legislative or regulatory constraints within the geographical sphere of her/his practice. Seldom will implementing 'zero discharge' be without substantial cost, yet this cost is often paid back over a relatively short period, with sometimes substantial savings thereafter. Measures that could move an enterprise towards 'zero discharge' are valuable and correspond essentially with measures to attain 'Cleaner Production' interpreted in the water sphere. Treatment of these details will be deferred until later in this chapter.

4.1.3 Hierarchy of water streams

Water streams in the water flow diagram will have varying qualities of water in them. Some processes intrinsically need high quality water, while others are served quite satisfactorily by water of a lower quality. The idea of an hierarchy of water streams is to assess whether the discharge from processes where high quality water is used is suitable input to processes where lower quality water supply is needed. Four levels of hierarchy are used: high stream quality, middle stream quality, low stream quality and bottom stream quality.

High stream quality is of at least potable quality and of even higher quality for boiler feed-water and some other processes. Middle stream quality, while of lower quality than 'high', is still adequate to imply sustainable plant operations and to ensure plant longevity. Cooling systems often operate on middle stream quality. (The exception is the use of seawater in once-through cooling systems, often in power stations.) Low quality streams are those used for such applications as dust control or which derive from processes such as mine de-watering from storm-water run-off, or from under-cover heavy vehicle wash down, which supplies water to contaminated water systems (see below). Bottom quality streams are those deriving from exposed contaminated water systems and from materials transport. Note that we have avoided quantifying the qualities of water that constitute the various streams. We seek presently to explain only the concept.

4.1.4 Geographical proximity

Enterprises within reasonably close geographical proximity might have water resources that can be shared. For example, wastewater available from one enterprise might be of adequate quality for the input water stream of a significant process in a nearby enterprise. Sometimes the possibilities can flow in both directions. When such shared arrangements are possible they can be very significant savers of water from primary water sources. Capital outlays can also be greatly reduced if nearby enterprises are able to share unit operations or water or wastewater treatment facilities.

4.2 UNPACKING CONCEPTS - IMPLEMENTING THE TOOLS

4.2.1 Cleaner production

We take some of the general tools of Cleaner Production and apply them solely to the water audit process. We assume that these matters will feature in general terms in the environmental policy statement of the auditee enterprise as part of its total environmental management strategy (if it exists).

4.2.1.1 Resource substitution

In our experience many enterprises use water of a higher quality than is needed for maintaining the integrity of some, or all, of their processes. In such cases the water auditor is encouraged to seek alternative sources of water. When scheme water is the primary source, other sources to be considered include ground water, storm water, seawater or wastewater from other enterprises geographically convenient. Clearly we need to measure the volumes of water used as a necessary pre-requisite for evaluating the sustainability of using alternative water sources.

4.2.1.2 Measurement

In itself measurement will not reduce water usage, but it is the essential prerequisite for both benchmarking and for the construction of the water flow diagram in the water audit process in addition to assessing sustainability.

4.2.1.3 Recycling and reuse of water and reduction of its use

The water flow or water management diagram provides the foundation upon which opportunities for water reuse or water recycling can be evaluated. 'Reuse' and 'recycling' are used rather interchangeably here, as what is recycling on one scale can be viewed as reuse on another, depending upon where the boundaries of unit operations are located. A key to establishing reuse and recycling possibilities is to evaluate when the quality of water discharged from one unit operation is suitable for recycling (maybe with some treatment) within the unit operation or when the discharge might be the input to another unit operation. The issues of water quality are discussed more fully below under 'Hierarchy'. We keep the two discussions apart for simplicity, although in practice in the field all factors are considered holisticly.

Water reuse or recycling will imply a reduction in water use from primary sources, yet the reduction of primary water use can be achieved by other strategies too. Among the strategies are separation of wastewater streams, efficiency improvements in one or more unit operations, changed practices, new technology and process automation.

4.2.1.4 Separation of waste water streams

A stream of wastewater can be rendered unsuitable either for recycling or for reuse by contaminants from a highly concentrated or toxic stream of even small flow rate being added to it. Separation of such streams might facilitate either the reuse or the recycling of some wastewater flows and thus enhance water use efficiency.

4.2.1.5 Efficiency

Water use efficiency will probably, but not necessarily, follow from improved process efficiency. In order for a water auditor to make credible suggestions regarding improved efficiency, he/she will need a keen understanding of the underlying principles of the operation. While the auditor's experience and careful listening might give rise to such understanding, in some cases it will be no match for an engineer's theoretical background coupled with specialised practical experience. We caution the general practitioner in water auditing to tread carefully. On the other hand, tradesmen and plant operators sometimes display insights that have been overlooked by professional engineers. Wisdom is not exclusive. Simple changes can sometimes cause significant improvements in water use efficiencies. Improving both process efficiency and water use efficiency will often involve the use of changed practices.

4.2.1.6 Changed practices

We select an apparently trivial example. If a process worker were using a hose to clean dry benign spillage from the floor, clearly a change would be required in his/her practice. The dry waste could simply be swept from the floor without the use of any water. There is also a possibility that the spillage could be returned to the process line. Changes in work practice require changes in human behaviour. Such changes are not always made easily. When they are made, there remains a strong probability that after a period of time the person or people involved will revert to earlier practices. While incentive schemes might help, the uncertain long-term viability of changes in human practices invites attention to a more permanent measure – new technology.

4.2.1.7 New Technology

We continue with the hose example that we introduced above . The installation of simple technology, either a nozzle on the process worker's hose (ie. high pressure low volume rather than low pressure high volume) or a flow restrictor in the tap, could lead to substantial water use reduction and increase in water use efficiency in some circumstances, without a change in behaviour of the worker. Even so, this simple response to the problem of water being wasted from the hose is not sufficient. A far more permanent approach would be to install new technology (or to modify existing plant), which doesn't allow spillage to occur as a matter of routine. Our simple example leaves aside the complexities of many industrial processes. Especially when rates of change in processes are high, automation is often a necessary adjunct to the introduction of new technology or for improving the performance of existing plant.

4.2.1.8 Process automation

Automation removes the process or processes from constant human manipulation of settings. Thus regressive human behaviour following changed work practices ceases to be an issue. Computers and sensors are able to respond to changes in systems at speeds and with sensitivities unachievable by humans. Total process efficiencies are likely and water savings in particular will be permanent.

Automation and the introduction of new technologies can be expensive. Both invite evaluation of costs, which will be taken up later in this chapter.

4.2.2 Zero discharge

The arenas where ' zero discharge ' might be implemented under legislation or by regulation differ from jurisdiction to jurisdiction. Here we can only give the broadest hints to alert the reader to possibilities. In Australia, for example, most environmental protection agencies enforceably apply the concept to mining sites, usually with a prescribed storm frequency of one in ten years and a storm duration of 2 hours. All potential run-off from the site is to be intercepted and impounded on the site unless a storm reaches or exceeds the prescribed storm. Oil storage facilities and chemical storage facilities, which are very environmentaly sensitive, are required to comply with 'zero discharge' up to a storm frequency and duration of one in five hundred years and twenty four hours respectively. At the other end of the spectrum such a concept is not generally applied in Australia to most light industrial sites, or to farms as examples.

Even when storms of greater gravity than the prescribed storm occur at a site where 'zero discharge' concepts have been implemented, pollution to the surrounding areas is likely to be reduced substantially. Much of the pollution we expect with surface run-off is picked up in the early stages of a storm. There is a high likelihood that this will be intercepted, and that later overflow discharges from the storm will be relatively clean.

4.2.3 Hierarchy of water streams

Unit operations using high quality water streams are evaluated first for the potential of their discharge streams to be used as inputs to processes with less demanding water quality requirements. Evaluations progress down the hierarchy until bottom quality streams are reached. This follows the natural order of cascading water quality as higher quality water streams leave their destined processes and are discharged usually as lower quality streams. These lower quality streams in turn are then assessed for suitability as inputs to unit operations where input water quality requirements are still less stringent.

4.2.3.1 High quality

In most cases process water streams provide water in a product, for example in a bottled beverage facility, or are evaporated. Sometimes it is cost effective to recover water discharged from high quality streams (example below). The recovered water can be applied as input water to wash-down facilities, used in water material transport systems, or importantly, used in cooling systems to make up for their system losses.

A particularly interesting example is boilers (and other steam raising cycles) which are usually 'blown down'. This means that water is removed from the boiler at regular intervals; sometimes it can be a continuous process. The blown-down water is then replaced by high quality water to maintain the water quality in the boiler at a satisfactory level, for the continual removal of steam gradually raises the level of total dissolved solids in the boiler water until the water quality endangers the boiler. Scale, a chemical deposition, can build up and impair boiler performance and consequently waste energy. In some cases, for example the build up of chlorides, chemical attack can take place on the boiler tubes or on the boiler shell. Approximately half of the blown-down high temperature water flashes to steam and is usually lost. The remainder is still of very good quality, usually much better than the feedwater to a demineralisation plant designed to supply feed-water to the boiler. Addition of the blown-down water after cooling to the normal stock of water for demineralisation, and then boiler feed, reduces demineralisation costs.

The destination of many discharges from unit operations employing potable water is the sewer. We reiterate that relatively good quality water should be considered for reuse, or recycling, before it is degraded by addition to the sewer, with its subsequent needs for costly treatment. Sewer water has high levels of pathogens and nutrients. Nevertheless treatment measures are available, which allow it to be used in some landscaping applications for example as well as other bottom quality applications.

4.2.3.2 Middle quality

Unit processes using middle quality water streams, in contrast with high quality streams, have the potential both to receive water from discharges from unit processes using high quality water and to supply water to unit processes using low or bottom quality water. Cooling water systems are typical users of middle quality water.

Most recirculating cooling water systems have the need for 'blow-down', in common with the boiler operations just discussed, especially where evaporation is involved as it is in cooling towers. This is necessary both to maintain the quality of the water and to protect the expensive plant in which the water functions as a coolant. Blowdown water from cooling systems is usually treated and a large proportion of it is recovered. The recovered water can then be recycled as input to the cooling water. If evaporation is the method of treating the blown-down water, then the treated water is demineralised and becomes a high quality resource. This method of treatment also implies a concentrated waste, which may be used to supply water for such applications as dust control or ash conditioning.

Cooling water systems can also receive water from other unit operations. Typically water waste from higher order streams is acceptable either in raw form or with treatment. Treated water from lower order streams can be acceptable in quality for input to unit operations requiring middle order quality water. For example, treated water from mine de-watering can be acceptable. Treated water from under-cover contaminated water systems is also worth evaluating as a source for cooling system water make-up. An example of a contaminated water system would be the separation system that removes oil and solids from water used to clean down heavy vehicles or machines. There are some dangers involved if residual oil levels are too high - oil films on heat transfer surfaces might interfere with the efficacy of the cooling water. Thus careful evaluation is essential.

4.2.3.3 Low quality

Most low quality water streams issue from unit operations. Few unit operations need low quality water as input for the intrinsic process of the unit. For example contaminated water systems process low quality water. But it is not the process which intrinsically needs the low quality water, but rather the low quality input water which needs the cleaning process. Many unit operations that receive low quality water can receive discharges from other processes, thus removing the need for their direct discharge off site. One example of a unit operation that can usefully receive waste low quality water (deriving probably from operations using middle quality inputs) is dust control.

As noted before, concentrate from cooling system blowdown, or its subsequent treatment, can be used cautiously for controlling dust from stockpiles of materials like coal or ores. Caution is required because impurities in the water used for dust control might be detrimental to the quality of the stockpiled material. An example of a detrimental effect is the presence of common salt on pulverised coal destined for burning in a boiler. The salt can cause premature wear on metal boiler surfaces that are subject to high temperatures.

4.2.3.4 Bottom quality

Storm water from 'dirty' sites will typically be bottom quality water. Sea-water might also be regarded so. Application of bottom quality water to materials transport must be qualified with the requirement that the quality of the water does not compromise the product in the transport system. When other applications are not obvious for bottom quality water, there are possibilities for its use after passing through a contaminated water system for removing debris and oils. The possibilities include groundwater recharge, landscaping irrigation and wash-down use. Table 4.1 summarises our discussion on the hierarchy of water streams. Note that the entries in the figure are illustrative rather than exhaustive.

4.2.3.5 Quantification

We have not quantified the qualities of water in our discussion on hierarchies of water streams. In a surprisingly large number of water audits we do not need to do so. As noted earlier, in many cases where quantitative data are required, the auditee enterprise will itself have measuring systems in place or it will have its own water quality laboratory. Detailed guidelines are available for water quality requirements for a large range of uses in Water Reuse (1992) to which we refer the interested reader and also in Chapter 8 of this book.

4.3 THE WATER MANAGEMENT STRATEGY

Reconsider now the Phase 3 - Water management strategy in the Figure 2.2. The summary there invites us to list the options, to evaluate each option and to construct a water management strategy. We now have the tools by which we can embark upon that process.

4.3.1 List the options

At this stage of the process generous latitude is appropriate for the water auditors' thinking. As in brainstorming, which is to be encouraged, uncensored ideas are to be welcomed. Only when creative thinking by all participants has been exhausted, should stringent critical faculties be brought to bear upon the options. Included in the options we would generally expect that water resources and resource substitution would feature, changes to internal processes in the audit domain would be considered, and options for the deployment of discharges internal to the domain and external to the domain would be canvassed. Probably the full range of tools discussed above would be used. The results of this process should be carefully recorded, so that they can be revisited easily. The linear process in Figure 2.2 in practice has some iterative features.

Table 4.1 Illustrative summary of the hierarchy of water streams, with some typical water sources, unit operations and discharges.

Unit operation uses	Typical source of water	Typical unit process	Typical quality of discharge
High quality water	Scheme water (treated) Fresh lake water (with treatment?) Distilled or ion exchange resin discharge water Rain water from a clean roof and gutter system (with filtering) Boiler blowdown water	Steam generation Food preparation Process water	High or middle
Middle quality water	Storm water from 'clean' areas Treated mine de-watering water Contaminated water system discharge	Cooling water systems Wash-down	Low or bottom
Low quality water	Mine de-watering water Uncovered contaminated water system	Material transport Dust control	Bottom
Bottom quality water	'First flush' storm water Sewage Concentrate from distillation water treatment Sea water, hyper-saline water	Once through cooling systems (Clean but saline) Material transport (with caveats) Landscaping (after treatment)	Bottom

4.3.2 Evaluate the options

At this stage rigorous constraints need to be brought to bear upon the options which have been identified previously. Evaluating the options is a multi-faceted task. Generally the factors are subsets of the broad categories of water issues, environmental issues, human issues, plant operations and financial issues.

4.3.2.1 Water issues

Typically water issues will be addressed by reality testing applied to external sources, internal recycling or reuse options and to discharge proposals. For example, if a proposal has been made to substitute say scheme water with bore water, is its quality suitable, is it available in sufficient quantities and in a sustainable way, is a license to withdraw the groundwater likely to be granted for the quantities needed? Similar questions are to be addressed to internal recycling and reuse proposals and to discharge proposals.

4.3.2.2 Environmental issues

Wider issues are at stake than merely water conservation and auditing. It is possible for a proposal to be positive in all respects from the perspective of water conservation, while impacting adversely on the environment. For example, if lower quality water were proposed as a substitute for existing cooling water in a conventional power station, this could adversely impact upon the thermodynamic efficiency of the power station. If in order to maintain plant operating life the temperature of the boiler needed to be lowered slightly, lower fuel use efficiency and impact upon emissions could follow. This implies that less effective use might be made of a non-renewable fuel and emission quality could deteriorate too.

4.3.2.3 Human issues

The introduction of automation in particular has consequences for employees and their families, and for local and regional economies and thus for other families. While the water auditor is not finally responsible for taking such decisions, he/she is responsible for alerting the client and the auditee to possible human consequences if the water auditor is aware of them.

4.3.2.4 Plant operations

Some options might imply that the plant, or some unit operations within it, needs to be shut down, or run at a lower level of production while modifications are being made. As has been noted before, changes of water usages within a plant can have consequent outcomes. While changes might result in a better product and a better overall performance for the auditee, this is not necessarily the case and the water auditor should evaluate options with this in mind.

4.3.2.5 Financial issues

A financial analysis should be undertaken for each of the options that survive other critiques. While quite intricate analyses could be undertaken of the costs and benefits of a water management strategy, we recommend that the water auditor confine her/himself to calculating simply payback periods of all the options. If more complex financial evaluations seem necessary, we recommend the services of a professional accountant. Factors to be considered include training costs for employees to operate new systems or technology, capital costs of implementing new water management strategies, differences in operating costs of the equivalent existing system, costs of failure of the water management strategy should that eventuate. Rates of return to be used in such calculations will be determined by the auditee enterprise.

The calculation of capital cost payback periods for each option is relatively straightforward. The cost of water savings per annum is calculated and divided into the capital cost of the option. More sophisticated methods of calculating a payback period will account for auditee internal rates of return and expected rates of inflation. Water auditing practitioners often divide payback periods into 'short term' (two or three years), 'medium term' (three to five years) and 'long term' (six years or over). Clearly the auditee will receive more readily options with a short or medium payback period.

The total capital cost of sets of options should be calculated. The auditee enterprise will be concerned not merely with payback periods, but also with its capacity to raise capital for proposed works and with the interest rates or other costs associated with raising the capital.

4.4 CONSTRUCT THE WATER MANAGEMENT STRATEGY

In response to the total evaluation process the proposed water management strategy should be presented with the most attractive options being given most weight, while still identifying less attractive options together with the reasons for them being less attractive.

4.5 REFERENCES

AS/NZ ISO 14004:1996 Environmental management systems – General guidelines on principles, systems and supporting techniques. Standards Australia and Standards New Zealand.

Dawson, M.D. (1997) Water audit guide. Case study - Wesfarmers CSBP Limited Kwinana Works. Independent study contract dissertation, School of Environmental Science, Murdoch University, Murdoch, Western Australia.

Schlafrig, J. (2000) Management strategies for water use. In Principles of water conservation and water auditing - Unit guide 2000, Murdoch University, Division of Science and Engineering, 157 pp.

United Nations Environment Programme (1996) Cleaner production - A training Resource package. 1st edition, UNEP Industry and Environment, France.

Water Reuse- Manual of practice SM-3 (1992) Water Pollution Control Federation, Alexandria, VA.

5

The water audit report

5.1 INTRODUCTION

In this chapter we move to phase four in Figure 2.2, the audit report. We will look firstly at the structure and contents of the report, delaying consideration of the 'process issues' represented in Figure 2.2 until the end of this chapter. The water audit report is the tangible end product of the water audit process itself. In the shortest space possible, the report is intended to convey to the client the essential issues, processes and water management strategy proposals that have arisen from the audit. Both implementation and management of implementation of the water management strategy lies outside of the scope of the audit process. The client, or the client and the auditee where the two are distinct from each other, controls both implementation and management of implementation of the water management strategy following hand-over of the report. In order to write a report, we think it is helpful to have a 'template' outline of contents for the auditor to follow. Occasions might arise when this template needs modification, but a basic outline of a water audit report helps ensure critical items are not overlooked.

We are aware that both the documents ISO 14010:1996 and ISO 14011:1996 provide contents lists for environmental audit reports. The lists are essentially the same, with that in ISO 14010:1996 being a general list and that in ISO 14011:1996 containing a small amount of additional reference to Environmental Management System audits. We will focus on the list in ISO 14010:1996. This list will provide useful input to us for our water audit report. Even so, the water audit process is sufficiently distinctive that we will need to extend significantly the list from ISO 14010:1996. Schlafrig (2000) provides a list specifically for a water audit, from which we will draw selectively.

5.2 CONTENTS

5.2.1 ISO 14010 contents

The non-exclusive contents list suggested in ISO 14010:1996 for environmental audit reports is reproduced below.

(1) Client and auditee identification,
(2) Scope and objectives of audit (agreed)
(3) Criteria (agreed)
(4) Dates of audit conduct and period of audit
(5) Audit team identification
(6) Auditee representatives identification
(7) Confidentiality issues
(8) Report distribution list
(9) Audit process summary
(10) Conclusions

We note that these headings include several, eg. (4,5,6 & 8), that appear to be minor to appear in a list alongside weightier issues like (9 & 10) for example. While we understand the need for flexibility for various situations, we find the list inadequate for a guide for either students or water auditors. Therefore we have constructed a contents list of our own, which we think distributes the weight of the contents more evenly between headings. The proposed 'contents' list is given below.

5.2.2 Water audit report contents list

(1) Title page
(2) Participants
(3) Executive summary
(4) The brief
(5) Introduction
(6) Description of site
(7) Plans for further development
(8) Water sources and sinks (general)
(9) Each water source (specific)

(10) Recycled and re-used water
(11) Each water sink (specific)
(12) Present water use and flow diagram(s) (including flow measurement methods)
(13) Raw options for improved water use efficiency
(14) Evaluation of raw options
(15) Water management strategy (including proposed water flow diagram(s))
(16) Water audit process review
(17) Conclusion

Each of these headings will be receive separate comment below. Note that all of the headings of the ISO list are included, though several might be buried under one new heading. For example the new heading 'Title page' includes (1,4,5,6,7 & 8) of the ISO list within its span.

5.3 DOCUMENT DETAILS

5.3.1 Title page

The title page would normally be of the form:

Water audit of [auditee name]'s [locality name] [audit domain name] [State] [Country]

Conducted by

[**Lead auditor name**]
of
[**Auditor business name**]

[for
Client name if not the same as auditee]

Period of conduct of on site components of the audit -
[dates defining period of audit]

Report completed on -
[Date of completion of audit report]

The bottom of the reverse side of the title page could have a copyright claim if necessary.

5.3.2 Participants

The following participants would be identified regardless of earlier appearance on the title page: client, auditee, audit team members, auditee's representatives and other significant participants like unit operation managers, union representatives or others who contribute significantly to the water audit's conduct. Acknowledgments are appropriate here. Confidentiality issues would be discussed and the consequent distribution list for dissemination of the audit report would be set down.

5.3.3 The executive summary

In the space of one page preferably, or not more than two, the aim of the audit and the essential financial savings to the enterprise, after the expiry of a defined payback period, should be identified. Water use savings should be identified, together with the essential implications of these savings for the future of the audit domain. Capital requirements for implementing the water management strategy are to be identified. Benefits in terms of water conservation and the enterprise's environmental management system are identified. Where criteria appear in the brief, the pertinent audit results should be compared with the criteria, with any associated implications.

The executive summary deserves to be the most carefully written part of the water audit report.

5.3.4 The brief

The brief is included here so that the reader of the total water audit report has the starting point conditions of the audit process and is left in no doubt as to the scope and objectives.

5.3.5 Introduction

The introduction includes comment on previous water audits, the brief and reasons for the current audit. Criteria, if relevant, are discussed, together with the reasons for the criteria being raised at this time. A brief evaluation might be made of the water management strategy as implemented from a previous audit. Changes of water resource availability, either physical or legal are very important to identify. Publications associated with water resource availability are to be mined for pertinent data. Changes in water use on the site, both historical and projected, form an important backdrop for the audit. In addition, the general economic background and the particular financial background for the water audit domain are of interest and could impact on decisions to implement the recommended water management strategy.

5.3.6 Description of site

This section only needs inclusion if the site description is too complex for adequate coverage in the introduction. Typical content would include the site principal usage, noting the interactions of this usage with the location, carefully set in the context of water availability or environmental sensitivity to discharges. The arrangement of buildings and operations on the site is of particular interest, especially if the site is subject to zero discharge provisions. Services other than water are of potential importance to the total water audit.

5.3.7 Plans for further development

This section, too, only needs inclusion if the future development plans for the water audit domain are sufficiently complex to be inappropriate to include totally in the introduction.

5.3.8 Water sources and sinks

All water sources and sinks could be included in this one section in a water audit of modest complexity. Typical sources and sinks were identified in Chapter 3 under 'Issues relating to input water from various sources' and 'Issues relating to water discharged to various sinks'. The purpose of this section of the water audit report is to draw attention to available water sources and sinks. In the case of sources, we evaluate their quality, assess the quantities available on a sustainable basis, note regulatory or legislative controls and evaluate the sources' potentials to supply water to the audit domain. Costs of water from these sources are also sought. In the case of sinks we evaluate the quality and quantity of wastewater that can be sustainably discharged to the various sinks again taking account of legislative or regulatory constraints and costs associated with the discharges.

5.3.9 Each water source and sink

Where particular sources or sinks have complexities, which make for clumsy inclusion in a general sources and sinks section, those sources or sinks may be elevated to occupy a section of their own.

5.3.11 Present water use and flow diagram

From the identification of unit operations, the identification of water flow streams and the flow measurements taken in the water audit domain, a water flow diagram is constructed of the situation as found. The flow measurement instrumentation and accuracy should be discussed here too. The water flow diagram is constructed as discussed in Chapter 3. This is the appropriate place to discuss flow measurement, quality measurement and instrumentation.

5.3.12 Recycled and re-used water

Opportunities for recycling or reuse of water from the water audit domain or from offsite are identified and evaluated in terms of quality and quantities. If treatment for the water seems necessary, the possibilities should be considered in this section.

5.3.13 Raw options for improved water use efficiency

The contents of this section of the report draw upon the outcomes of the process discussed within the body of Chapter 4, 'Management strategy', 'List the options'. Because of the brainstorming nature of the process outlined there, some preliminary sifting of the options is necessary before inclusion in the report. We recommend that less rather than more sifting is appropriate at this stage. The criteria for such preliminary sifting are not easy to articulate. We suggest that what appear heuristically to be the most viable options are included in the body of the report, with the remainder being recorded in an appendix. No quantitative analysis is warranted in this section. The prime purpose of this section in the report is to provide the raw material for evaluation of options to be readily revisited, should the evaluations of the next section become out-dated between the final presentation of the report and the start of implementation of the preferred water management strategy. For example a shift in cost of raw materials or labour might move the balance from the recommended options of the following sections to others not recommended there.

5.3.14 Evaluation of raw options

Here we record the processes used and the outcomes obtained from 'Evaluate the options' described in the body of Chapter 4, management strategy. Those options that are favoured at the end of the evaluation process should be clearly identified and weightings identified for the water issues, environmental issues, human issues, plant operations and financial issues, which have led to the auditor favouring some over others.

5.3.15 Water management strategy (including proposed water flow diagram(s))

The favoured options of the previous section are now drawn together in a unitary way to constitute the total water management strategy. (See Chapter 4 too.) This section summarises and integrates the outcomes of the previous section. If changes are implied to the water flow diagram(s) under ' Present water use and flow diagram', the new water flow diagram(s) are usefully presented here. The financial evaluations are presented last and normally in terms of payback period. The amount of capital needed to be raised for the water management strategy and the methods for raising it are discussed.

5.3.16 Water audit process review

In the course of conducting a substantial audit the auditor will experience unexpected and sometimes unhelpful events, which do not contribute to the smoothness of the audit process. Reflecting upon these events enables the auditor to identify personal changes, and process changes, that might ease the progress of future audits on this or any site. In the audit report itself only changes which constitute recommendations to the client or to the auditee have a place. Positive experiences are likely on the way through the audit process too. It is appropriate to affirm the auditee enterprise or to thank those who have been helpful. An alternative to this section, especially for small audits, is 'Acknowledgments'.

5.3.17 Conclusions

This section is particularly valuable where the audit brief has included criteria (quality and quantity of discharges to a sewer for example) against which audit results are to be compared. Otherwise the section 'Water management strategy' constitutes a conclusion section. The brief itself can imply what is suitable for the conclusion. While summarising material could usefully be gathered, the 'Executive summary' is also summarising in nature. In some instances the 'conclusions' section might be omitted.

5.3.18 Process issues

Again we emphasise that the notional linear process displayed in Figure 2.2 has iterative aspects to it as it is actually implemented. The author of the report would normally liaise extensively with personnel of the auditee enterprise while writing the report. The summary and recommendations are titled 'Executive summary' and either 'Water management strategy' or 'Conclusions' or both in the contents list we have suggested. Though these sections will probably be written last, they themselves will likely go through several revisions as a result of consultation with senior staff. We suggest that a *draft* report be presented to representatives of the auditee (or client and auditee, when different) in seminar style. This enables all significant stakeholders in the outcome of the audit to make suggestions and to respond to counter suggestions in an open dialectical process. Such a process also protects the water auditor from unwittingly making spurious recommendations. Only when modifications from the seminar process are integrated into the report, or not integrated with good reasons, should the final report be presented to the auditee's representative (or the client's and auditee's representatives when the two are different). An example of a complete small, but illustrative, water auditing report is given in Chapter 23.

We have now progressed through Figure 2.2 'Water audit process'. There remain outstanding issues that have been sidelined so far for the sake of continuity in the text; these are the means of measuring water flows, water quality and legal issues that might set criteria of interest to the water auditor in a particular legal jurisdiction. We will summarise and integrate our progress thus far in Chapter 6, in Chapters 7 and 8 deal with 'Instrumentation and flow measurement' and 'Water quality' respectively, and in Chapter 9 with 'Environmental legislation – impact on water'.

5.4 REFERENCES

AS/NZS ISO 14010:1996 *Guidelines for environmental auditing – General principles.* Standards Australia and Standards New Zealand.

AS/NZS ISO 14011:1996 *Guidelines for environmental auditing – Audit procedures – Auditing of environmental management systems.* Standards Australia and Standards New Zealand.

Schlafrig, J. (2000) Procedures: Part 2 - The audit report - in *Principles of water conservation and water auditing - Unit guide 2000*, Murdoch University, Division of Science and Engineering, 157 pp.

6

Water auditing and environmental auditing revisited

Following the detailed examination of the water auditing process in Chapters 3, 4 and 5, we will now further refine and summarise our view of the relationship between water auditing and environmental auditing. We will employ several Venn diagrams to illustrate our conclusions.

Our starting point in Chapter 2 was that water auditing appears to be a subset of environmental auditing as water is a component only of the total environment. We might represent this in Venn diagrams as in Figure 6.1.

6.1 A BROADER VIEW OF WATER AUDITING

While Figure 6.1 is indeed true, it is not a comprehensive representation of the truth and, by itself, is misleading. This is illustrated in the Table 6.1 below, where a tabulated comparison is made of the environmental management system and the water auditing process. Notice that in the water auditing process column, the items 'Water conservation…' and 'Implementation' do not appear in the Figure 2.2. This is because Figure 2.2 represents the water auditing process from the point of view of the water auditor, rather than the auditee enterprise or the client, for whom the process is broader.

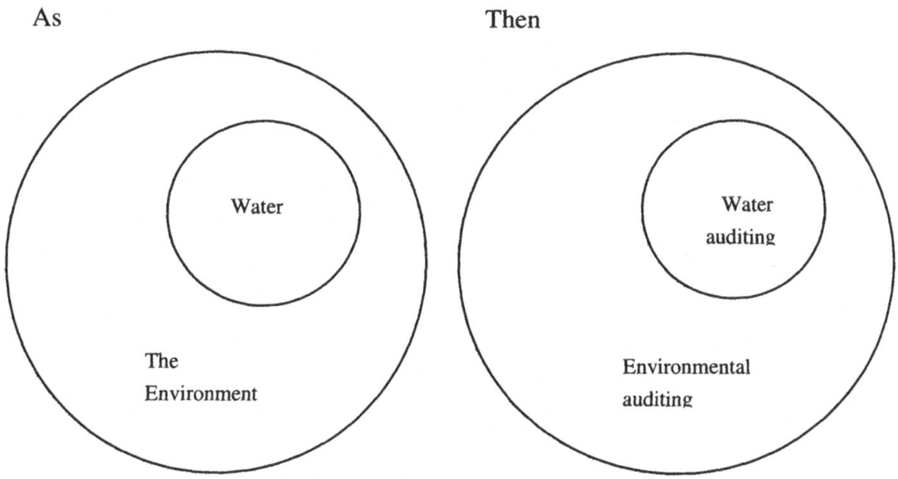

Figure 6.1 First glance interpretation of the qualitative relationship between environmental auditing and water auditing.

Table 6.1 A comparison of the environmental management system model with the water auditing process, taking the broader view of water auditing that client and auditee might have compared with the water audit practitioner.

The E.M.S. model	The water audit process (holistic)
Commitment and policy	Commitment to water conservation, water resource substitution, reuse and recycling, financial evaluation.
Planning	Figure 2.3 phases 3 and 4.
Implementation	Implementation. (Outside of process as viewed by the auditor)
Measurement and evaluation	Figure 2.3 phases 2 and 3.
Review and improvement	Figure 2.3 phases 3 and 4.

Table 6.1 demonstrates the significant parallels in the environmental management system model and the broader perspective on the water auditing process. The environmental management system will put in place criteria for an

environmental audit to check against, and this is true for at least the wastewater audit part of water auditing. Yet the total water audit is concerned with achieving commitment to water conservation, water resource substitution, reuse and recycling and positive financial outcomes for the auditee where possible.

6.2 SYSTEMS AND PROCESSES

From the point of view of systems and processes, Figure 6.2 demonstrates the relationship between environmental management systems, environmental auditing and water auditing. Figure 6.3 shows the relationship between pollution control, basic resource conservation and water conservation, the basic value underlying water auditing. We believe that environmental auditing is weighted towards pollution control, while water auditing is weighted towards basic resource conservation. Thus environmental auditing tilts towards conformity with criteria related to pollution control, while water auditing tilts towards water conservation apart from pre-determined criteria. These observations are matters of emphasis only and should not be carelessly simplified.

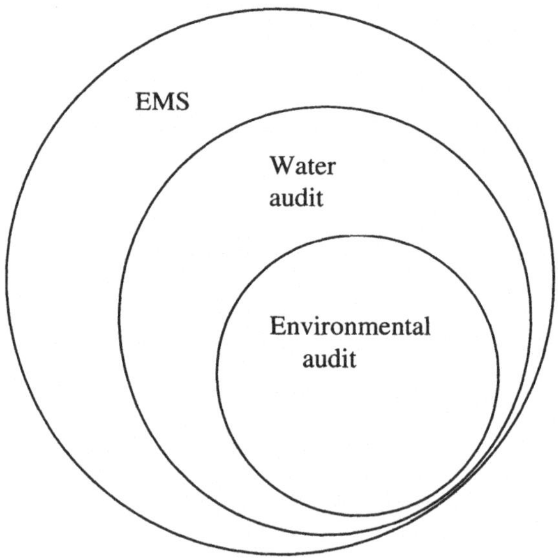

Figure 6.2 The qualitative relationship between environmental management systems, environmental audits and water auditing.

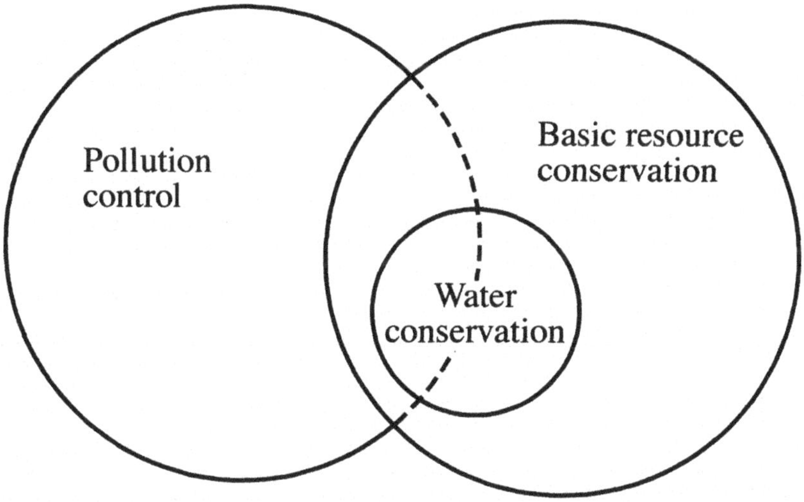

Figure 6.3 The qualitative relationship between basic resource conservation, pollution control and water conservation.

The three diagrams of Figures 6.1, 6.2 and 6.3, together with the Table 6.1, give a qualitative picture of the relationships between environmental auditing, water auditing and water conservation. The discussion in Chapter 2 has thus been extended and refined as foreshadowed both there and in the introduction to Chapter 3.

7

Instrumentation and flow measurement

7.1 INTRODUCTION

In this chapter we address the first of the two outstanding matters from the previous chapters three through to five, in which we have worked through the 'Water audit process' as visualised in Figure 2.2. Instrumentation and flow measurement is the foundation and source for quantifying the water flow streams in water flow diagrams. In this chapter we will present sufficient information to assist the water auditor to make flow measurements, without undertaking the physical derivations which properly belong to fluid mechanics texts (eg. Robertson & Crowe (1993) and Daugherty *et al.* (1989)) and without supplying extensive details that are found in handbooks (eg. Miller (1993) and Cheremisinoff & Cheremisinoff (1988)). The water auditor will need to measure flows in three physical arrangements. In general terms we can describe them as open channel flows, flows which do not fill an enclosing pipe and flows which do fill an enclosing pipe. Occasionally a water auditor might need to estimate run-off flows over soils. In this case we refer the water auditor to a surface water hydrology book, eg. Ward & Elliott

(1995). We could classify measuring methods based on the type of primary measuring element too. Overall, flow measurement methods can be either direct or indirect.

Direct methods involve the primary measurement of either a mass or a volume of water over a measured span of time. An example is filling a container with water of known volume and measuring the period of time to fill it with a stop-watch. (This is a surprisingly versatile method, which we will re-visit below). Indirect flow measurement methods are those that measure some physical parameter, which is a function of flow rate, where the function is determined by applying physical laws. For example, measuring the height of water in a straight open channel of known geometry, slope and surface roughness would be sufficient to deduce the flow rate from fluid mechanical principles, with a small error associated with the deduction. The error would normally be dealt with by means of a correcting coefficient, determined empirically. This example illustrates the general fact that while physical reasoning gives good insight into the overall characteristics of the indirect measuring instrument, nevertheless the precise value of a numerical correction needs to be found by calibration. The water auditor would usually, but not always, use an instrument that is pre-calibrated directly or indirectly against a national or international standard. In the bulk of this chapter we will focus on indirect measuring methods in each of the three physical arrangements mentioned above. A short section late in the chapter will return to the topic of direct flow measurement.

In addition to considering the methods employed by flow measuring devices or their classification, there are several factors worth considering when selecting a flow-measuring instrument. The flow-rate is rather obvious, but the flow's variability is important too. What timescale of variability do we want to detect? An alternative way of thinking about this matter is to ask what is the integration timescale over which we will accept a mean value of the variable flow rate? The characteristics of the water are important to consider too. As we will see later in this chapter, some flow measuring instruments require low quantities of small solid impurities; others require some fluid salinity or, more correctly, conductivity. The type of pipe and its accessibility will also bear upon a selection, as will the pressure of the fluid whose flow rate is to be determined. Finally, and importantly, the cost of the measuring devices will have a large impact upon the choice of measuring device for a particular situation. Having flagged these selection issues, we proceed to examine types of measuring devices suitable for use in each of open channel flows, pipes partially filled with water and pipes wholly filled with water.

7.2 FLOW IN OPEN CHANNELS

Often we measure flow rates in open channel flows by means of a weir. A weir is a vertical obstruction that is placed transversely across the whole of the path of the flow. The height of water above the crest of the weir is related to flow rate, so that if the water level above the weir's crest is measured the flow rate can be calculated.

There are several different weirs in wide use. They are the rectangular, trapezoidal (Cipolleti) and the triangular (V-notch) weirs. Figure 7.1 illustrates the three types, with some of the required constraints on their dimensions. We will address each type in turn.

7.2.1 The rectangular weir

Rectangular weirs are usually used where the flow rate is reasonably steady, for a rectangular weir is not accurate at measuring flow rates that are much lower than those for which the weir has been selected. The relationship between flow-rate Q (m^3s^{-1}) and the height of water above the crest of the weir H (m) is

$$Q = \frac{2}{3} CL\sqrt{2gH^3} \tag{7.1}$$

where	C (dimensionless)	is a discharge coefficient, of order (1),
	L (m)	is the breadth of the weir, see Figure 7.1,
and	g (ms^{-2})	is the acceleration due to gravity.

The crest of the weir needs to be 'sharp', where the meaning of a sharp crest is shown in Figure 7.2, which also illustrates constraints on the location where H is to be measured as well as the requirements for settling length in a straight channel upstream of the weir. A typical value of C is 0.58, but the precise value should be determined by calibration of individual weirs. When the crest of a 'rectangular' weir extends across the whole of the channel width, the coefficient C will assume a larger value (typically 0.62) than for the arrangement shown in Figure 7.1, which results in the nappe contracting laterally. Again the precise value should be determined by calibration of individual weirs. (See Daugherty *et al.* (1989) for further details.)

7.2.2 The trapezoidal (Cipolletti) weir

The trapezoidal or Cipolletti weir (see Figure 7.1) approximately compensates for lateral contraction of the nappe in the arrangement of Figure 7.1. Thus the coefficient is similar to that of a weir with a crest fully spanning the channel, while equation 7.1 continues to be appropriate for the discharge.

7.2.3 The triangular weir

The chief advantage of the triangular weir is that it can measure accurately flows that are considerably smaller than the maximum flow rate for which the weir is selected. The relationship between flow-rate Q (m^3s^{-1}) and the height of water above the crest of the weir H (m) is

$$Q = \frac{8}{15} C \tan(\theta/2)\sqrt{2gH^5} \tag{7.2}$$

where θ is the angle of the V-notch. A value of $\theta = 90^\circ$ is common,
and C typically lies between 0.58 and 0.61.

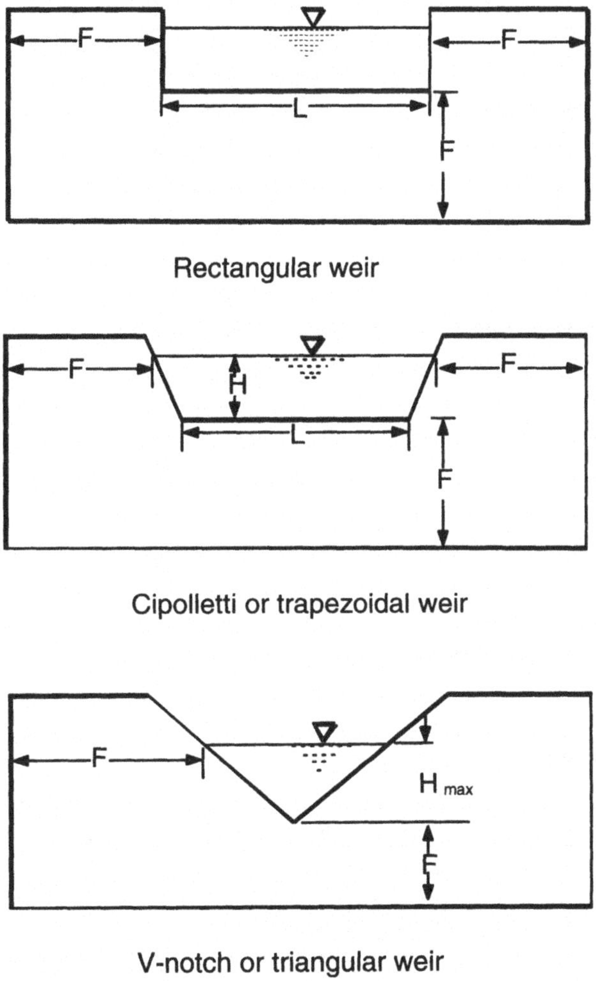

Figure 7.1 Configurations of rectangular, trapezoidal and triangular weirs

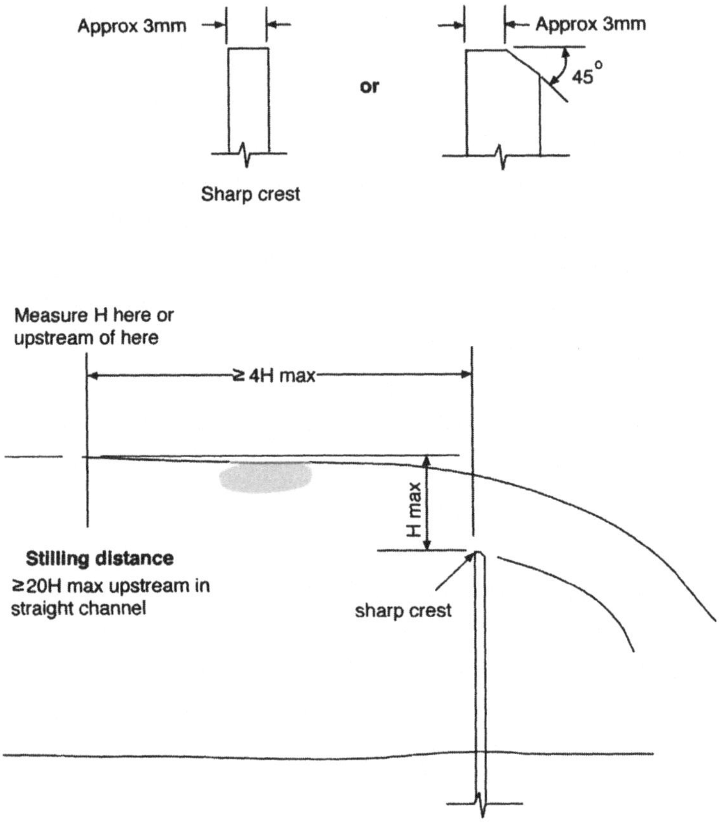

Figure 7.2 Sharp crested weir and measurement constraints.

7.2.4 Flumes

A flume is a measuring device that causes minimal obstruction to the flow. In weirs the obstruction to flow is in the plane normal to the flow direction in the channel. When there is insufficient channel depth to meet the requirement for weir height the obstruction can extend minimally from the channel walls and extend over a horizontal region on the bottom of the channel and/or over the sidewalls of the channel. This constitutes a flume. Standard flumes (eg. Palmer-Bowlus, see Figure 7.3) have been developed. Constructing the flumes based on these geometries enable flow to be determined from water level upstream of the flume. (See Miller (1993) for further details.)

7.2.5 Other methods

Velocity at selected points across an open channel flow can be measured using a variety of instruments: turbine, ultrasonic and pitot tube. These devices will be treated below in more detail. The total flow can be calculated by integrating the velocity profile across the flow cross-section. We do not recommend these methods unless there are compelling reasons to use them.

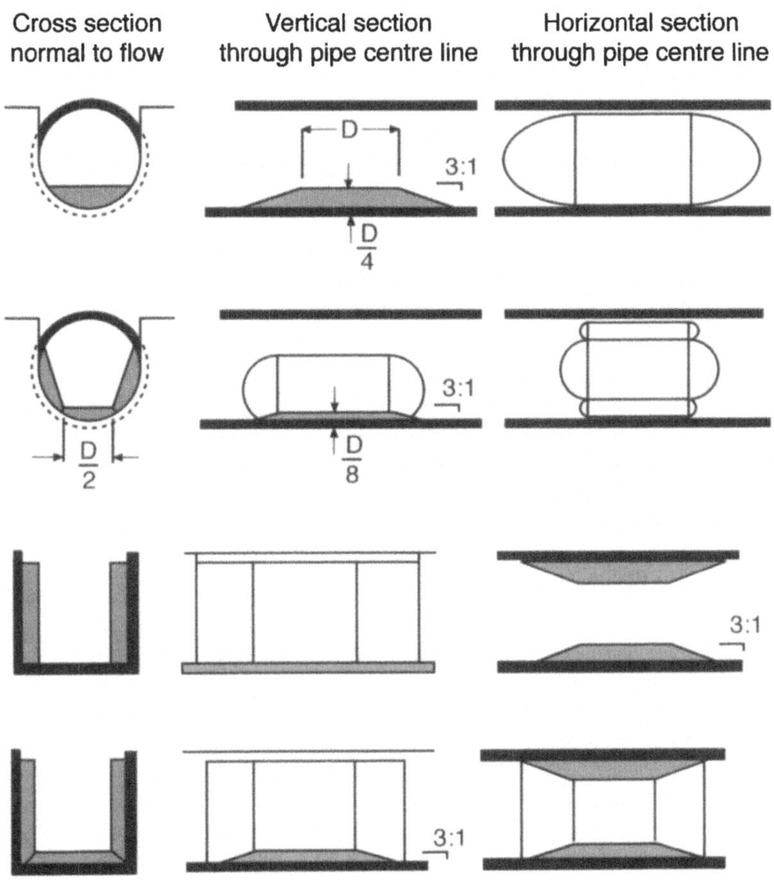

Figure 7.3 Some typical flume configurations.

7.2.6 Improvisation

A cork or piece of foam travels with the stream in an open channel flow or unfilled pipe. We can measure the time the float travels a certain distance. The average flow velocity in unfilled pipe is about 0.8 the velocity of the float. The accuracy of this method may not be better than ±10%, but can be appropriate on occasion.

7.2.7 Accuracy

Typically the accuracy of weirs is ±4%, but with great care the accuracy can be increased to ±1.5%.

7.3 FLOW IN PIPES PARTIALLY FILLED WITH WATER

7.3.1 Flumes

Flumes, as discussed above, can also be used to measure the flow in partially filled pipes (see, too, Figure 7.3). Also see Miller (1993) for further details.

7.3.2 A simple geometric method

Figure 7.4 illustrates a simple geometric method for determining the flow from a partially (or wholly) filled pipe. This method is especially practical when there is little head producing the discharge. The discharge Q ($\mathrm{Lmin^{-1}}$) is related to the horizontal distance measured as shown in the figure:

$$Q = 0.289 A_w X_o \qquad (7.3)$$

where A_w is the cross sectional area of the fluid in the pipe,
and X_o is the horizontal reach of the stream as shown in Figure 7.4.

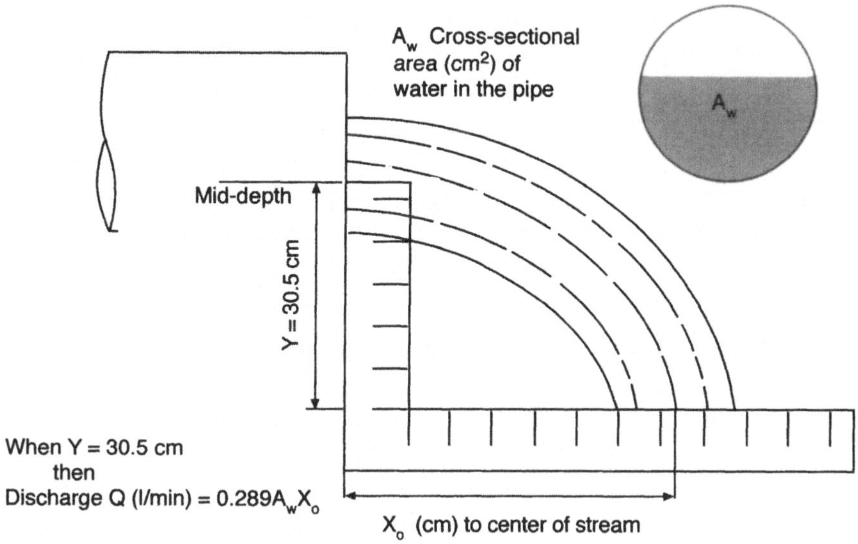

Figure 7.4 Measurement of flow flow a partially filled pipe – a simple method.

The cross sectional area of the fluid in the pipe can be determined approximately from Figure 7.5. Alternatively the cross sectional area of the water can be found by simple geometry to be

$$A_w = \frac{D}{2}\;\frac{D}{2}\theta - (D - 2d)\sin\theta \qquad (7.4)$$

where $$\theta = \text{arcos } 1 - \frac{2d}{D} \qquad (7.5)$$

when $0 \leq \theta \leq \pi$ *and* D and d are defined in Figure 7.5.

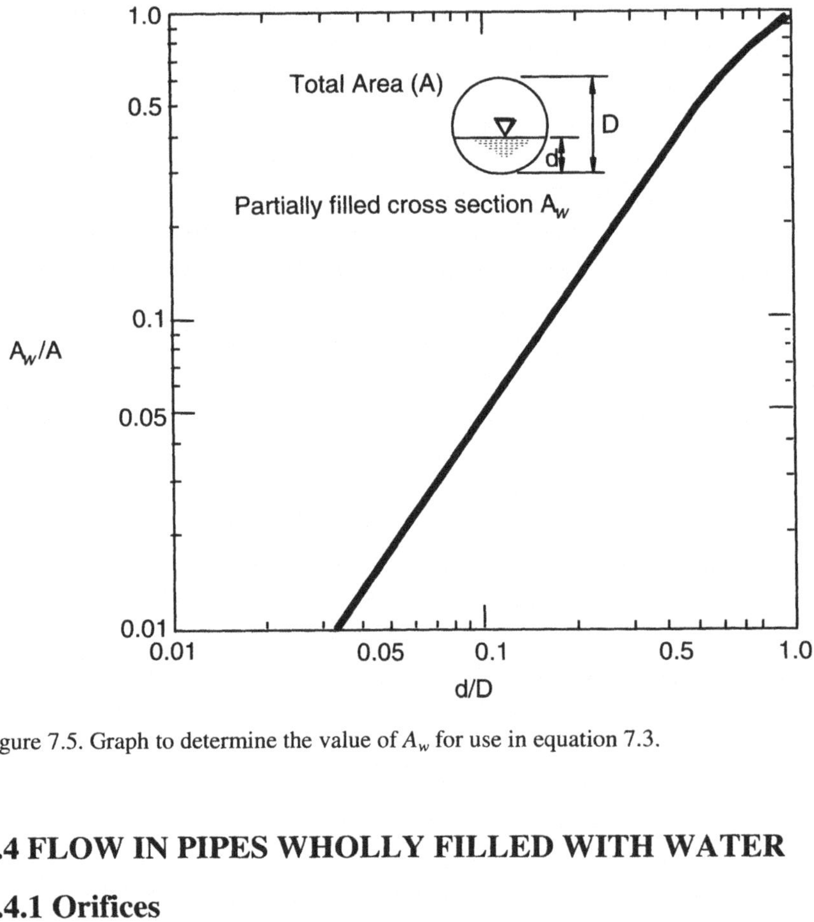

Figure 7.5. Graph to determine the value of A_w for use in equation 7.3.

7.4 FLOW IN PIPES WHOLLY FILLED WITH WATER

7.4.1 Orifices

A common means of measuring flow in pipes, where the water occupies the full cross-sectional area of the pipe, is by inserting a restriction to the flow and measuring the pressure drop across the restriction. A relatively inexpensive way is to install an orifice between pipe flanges. See Figure 7.6, which illustrates a range of orifice types.

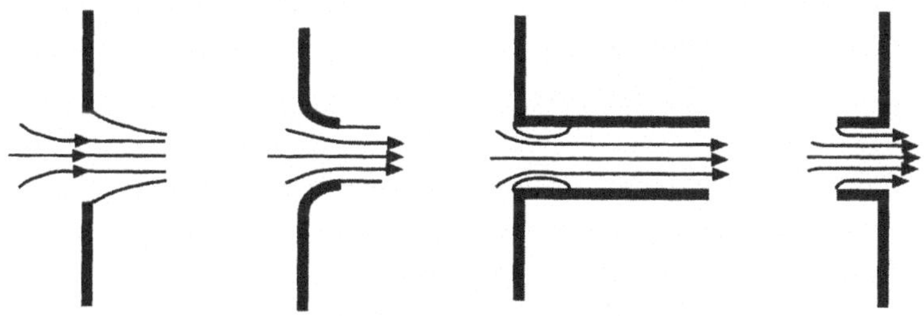

Figure 7.6 Various orifice like flow measuring devices. (C= 0.61, 0.98, 0.80, 0.51 respectively).

The flow rate (Q) is related to the head loss (ΔH) across the orifice by

$$Q = CAK\sqrt{2g\Delta H} \tag{7.6}$$

where C is the discharge coefficient

A is the cross-sectional area of pipe

$K = \sqrt{1/\left\{1 - (A_o/A)^2\right\}}$

A_o is the area of the throat

$\Delta H = (p_1 - p_2)/\gamma$ when pressure is measured rather than differential head

p_1 and p_2 are the upstream and downstream pressure measurements

and γ is the specific weight of the fluid

Pressure difference between the upstream and downstream ports can be measured using a manometer or differential pressure transducer. The discharge coefficient (C) is listed in Figure 7.6 for a number of orifices.

Though common and inexpensive, orifices have the disadvantage that there is a large permanent pressure loss. Thus the measuring device can substantially reduce the flow-rate in the line being measured. A Venturi meter overcomes this disadvantage significantly.

7.4.2 Venturi meter

A Venturi flow meter is shown diagrammatically in Figure 7.7. The relationship between the head loss from upstream to the throat of the venturi tube is given by

$$Q = CA_t \sqrt{\frac{2g\Delta H}{1-(D_t/D_u)^4}} \quad (7.7)$$

where C is a coefficient usually between 0.98 and 1.0
A_t is the throat area
D_t is the diameter of the throat
D_u is the upstream pipe diameter
and other symbols have the meanings given previously.

Venturi tubes have the disadvantage of being more expensive than orifices, though the relatively small modification Venturi's make to the measured flow often more than compensates for the cost. Both orifice plates and Venturi tubes have the added disadvantage of requiring a pipe to be cut, flanges fitted and the device installed. Thus the process using the water flow needs to be shut down, or an alternative water supply provided.

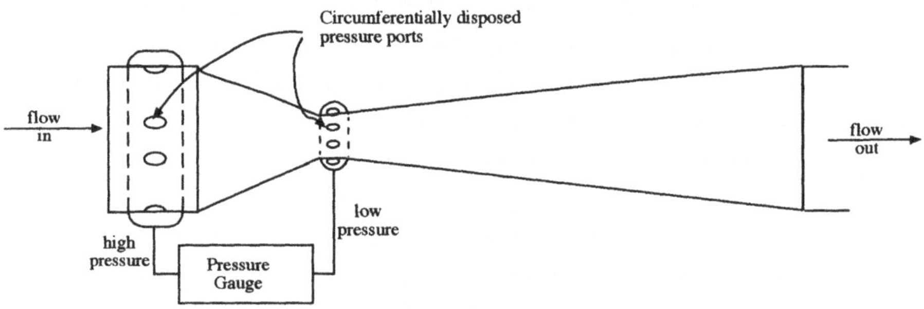

Figure7.7 Diagrammatic representation of a Venturi meter.

7.4.3 Pitot tube

The principle of the pitot tube is illustrated in Figure 7.8. It has the advantage of not interfering with the flow to the extent of orifices, but somewhat more so than for Venturi tubes. The installation process can be less onerous. We do not recommend the pitot tube in its classical form of a single tube with one stagnation point for use by water auditors. This is because it requires a traverse taking several velocity readings, which are then integrated across the pipe to give rise to a flow rate. We include the pitot tube because there are now manufactured pitot tubes with several stagnation ports spanning the diameter of the pipe. Associated electronics interpret the readings and provide an instantaneous flow value for the whole-of-pipe flow. The accuracy is claimed to be approximately ±0.5%. For completeness we provide the relationship between velocity and differential pressure between the dynamic and static pressure ports:

$$u = C\sqrt{2g\{(p_1 - p_2)/\gamma\}} \tag{7.8}$$

where C is a coefficient typically between 0.98 and 1.0
and the other variables have the same meanings as above.

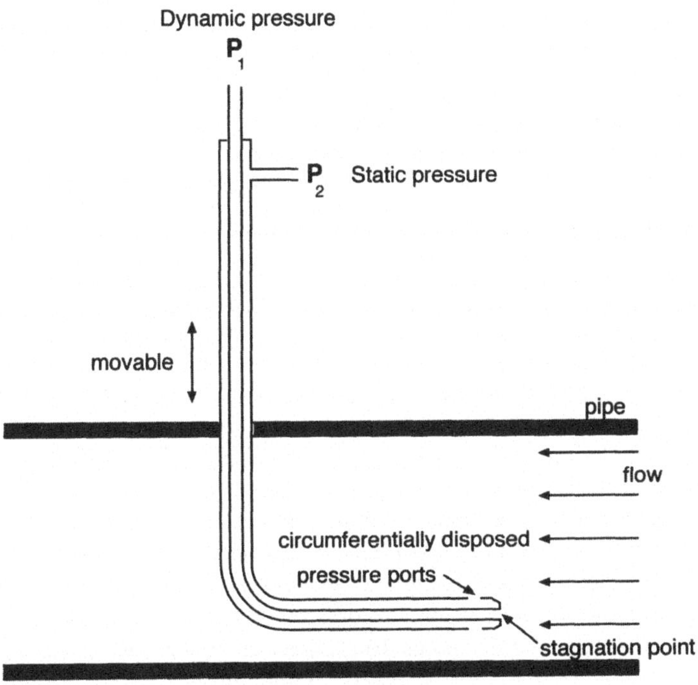

Figure 7.8 Diagrammatic representation of a pitot tube.

7.4.4 Fouling

All orifices, Venturi tubes and pitot tubes have pressure ports or tappings associated with them. Fouling of these tappings is inevitable over time and can give rise to substantial errors in the flows measured. Thus it is essential that the tappings be cleaned form time to time. This is a major disadvantage if the line has to be shut off and the associated process disturbed or stopped.

7.4.5 Rotary flow meters or turbines

A rotary flow meter or turbine is a multiple bladed 'propeller' installed in the line in which we wish to measure flow. By means of an electromagnetic pick-up located in the meter housing, the rotational rate of the wheel expressed as an electrical voltage or as a count of revolutions is calibrated against volume flow rate. Commonly, digital electronics associated with the meter enable an instantaneous reading of flow rate or other statistics such as total volume of fluid past the meter since a selected time.

These meters are particularly sensitive to fouling and the calibration also tends to drift with time as the turbine bearings wear. On the other hand the major advantages of these meters include their relatively low cost, ability to measure low as well as high flow rates accurately, their suitability for water, non-acid chemical solutions or liquid chemicals.

7.4.6 Electromagnetic flow measurement

The basic principle of the electromagnetic flow measuring device is the same as that of an electrical generator, that is, a conductor moving in a magnetic field produces an electromagnetic force (voltage) over its length. In the case of a fluid, electrodes on diametrically opposed sides of the pipe sense the voltage induced in the fluid by magnets associated with the sensors in a non-intrusive unit installed in the pipe. (See Figure 7.9.) The flow is almost directly proportional to the voltage measured at the electrodes. The measuring unit does not interfere with the flow, has no pressure tappings and also has no moving parts. The devices can be obtained with linings resistant to specific chemicals or a range of chemicals. Data logging is easy either remotely or on site.

The cost of these devices is high, a disadvantage. Additionally, in order for the device to work a minimal amount of 'salt' or fluid conductivity is required; about 1000 μS/m or, in older units, 10 μmhos/cm. Scheme water in most of the world would exceed these requirements. Accuracy is high at typically ±0.5% over a wide range of flow rates or velocities.

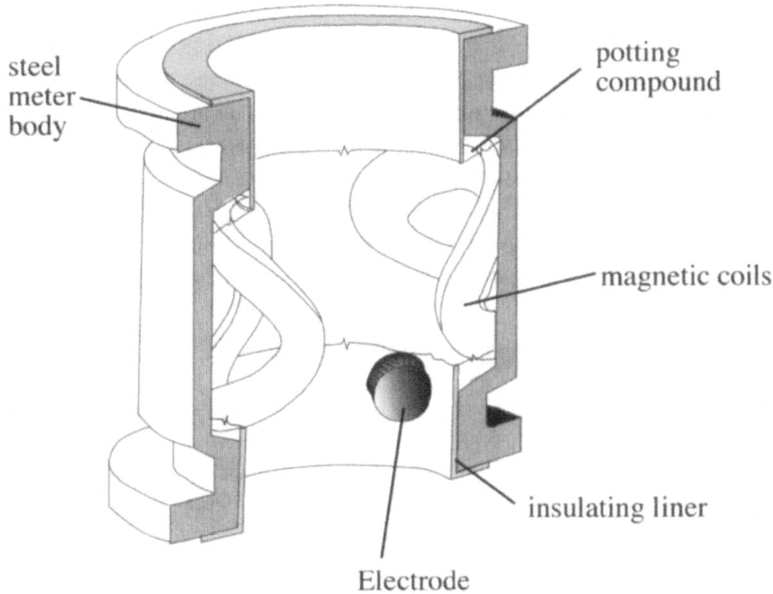

Figure 7.9 Diagrammatic representation of electromagnetic flow meter.

7.4.7 Acoustic Doppler velocimeters (ADVs)

The principle of these devices is that if an acoustic signal reflects off an object with a relative velocity to the source, the reflected signal shifts down in frequency if the object is moving away from the signal source or vice-versa. ADV probes rely upon small particles suspended in the fluid in order to measure the velocity of the fluid.

Two types of ADV might be of interest to water auditors. One can be clamped on to a pipe, so enabling non-intrusive flow measurement with no need to break the line to install the instrument. The second is a probe that measures velocity in a small zone near the acoustic sources and which is suitable for traversing in open channel flows or being deployed from a boat in a stream. These devices are quite expensive.

7.4.8 Positive displacement meters

Scheme water supply providers often meter water usage at the point of entry to houses or factories using a positive displacement meter with a digital readout. These meters are available in a variety of sizes. The user will find that it is cost effective to use the smallest sized meter that will cover the enterprise's actual water usage.

7.5 ACCURACY

Devices relying on differential pressure measurements, ie. orifices, Venturi meters, (classical) pitot tubes will generally give an accuracy of ±5%. With care Pitot tubes can be significantly better than this. Turbines are usually accurate to ±2%, electromagnetic flow meters to ±0.5% and ultrasonic flowmeters to approximately ±2% also. These figures are indicative only, and the expected accuracy of flow measuring devices used by a water auditor should always be determined from the manufacturer or from calibration trials.

7.6 DIRECT FLOW MEASUREMENT

Where flows discharge from an open channel, partially filled pipe or wholly filled pipe we have the opportunity to employ direct flow measurement. Essentially this involves filling a container to a calibrated level over a measured period of time, eg. using a bucket and stop watch. This method has the great advantage of low cost, speed and surprising accuracy. Let us investigate the accuracy with an example.

Suppose that 5.0 L of water is collected in a bucket in 51.2s. The calculated flow rate of the water entering the bucket is therefore 5.0 L / 50.2 s = 0.098 L/s.

The bucket had been calibrated to intervals of 0.2 litres. Assume that the quantity of water is accurate to plus or minus half the least scale division. The error attached to the water volume is therefore ±0.1 L, or a percentage error of ±2%.

The least scale division of the stopwatch is ±0.2 s, but we believe that the error due to reaction time overwhelms this source of error, without extensive practice with the stop-watch. An error of ±0.4 s , at both the start and the end of the time period seems reasonable. Therefore the time interval error is ± (0.4 + 0.4) s = ± 0.8 s or ± 1.6%.

The error for the calculated flow rate is the root mean square of the errors of the volume and the time, for division is involved. Thus the error in the flow rate is

$$\sqrt{2^2 + 1.6^2} = 2.6\%$$

Reinterpreting this percentage error as a flow results in 0.026 x 0.096 = 0.002 L/s.
The water flow rate is 0.098 ± 0.002 L/s or 0.098 L/s ± 2.6%.

Whatever the merits of the example selected, it is clear that remarkable accuracy is possible by the simplest means of direct flow measurement.

7.7 REFERENCES

Cheremisinoff, N.P. & Cheremisinoff P.N. 1988 *Flow measurement for engineers and scientists.* New York, M. Dekker, 392 p.

Daugherty, R.L., Franzini, J.B. & Finnemore, E.J. 1989 *Fluid mechanics with engineering applications SI Metric edition.*. McGraw-Hill, Singapore

Miller, R.W. 1993 *Flow measurement engineering handbook.* McGraw-Hill, New York.

Robertson, J.A. & Crowe, C.T. 1993 *Engineering Fluid Mechanics* Houghton.

Scott, R.W.W. (ed.) 1982 *Developments in Flow Measurement - 1.* Appl. Science Publisher, Englewood- Cliffs, N.J.

Shercliff, J.A. 1962 *Electromagnetic flow-measurement.* C.U.P., New York.

Ward, A.D. & Elliot, W.J. (Eds) 1995 *Environmental hydrology* . Lewis Publishers, Boca Raton, Fla.

8

Water Quality

8.1 WATER QUANTITY AND WATER QUALITY

Quantifying the amount of water use is naturally a primary aim of water auditing, because our objective in conducting water auditing is to reduce the amount of water use. Consideration of water quality is, however, equally as important as water quantity. The use of water is dependent on its quality. Very good quality water can be used for human consumption (drinking and cooking purposes), but also for other uses such as for agriculture (irrigating crops). On the other hand low quality water may only be used for limited purposes. The determination of water quality is therefore an important part of a water auditing process. For each stream for which water quantity is measured, there should ideally be a determination of its water quality.

In this chapter we will consider the parameters that determine water quality. We will then consider a classification of water into various grades based on just a few parameters. Though the classification is simplified it is based on established practices for purifying water, and will provide a general indication of quality of water.

An alternative classification of water that is related to its quality is based on what is called the beneficial use of the water. It is based on more rigorous studies of what plants can tolerate, for example, if the beneficial use of the water is for irrigating plants. Water quality criteria for various beneficial uses of water have been developed based on scientific investigation, and levels of water quality parameters established for these uses. We will consider these beneficial uses and their corresponding quality criteria.

Understanding water quality will inform water reuse possibilities following the completion of a water auditing exercise that includes both water quantity and water quality. It will fulfil the broader objective of such an auditing, which is better use of water and hence achieving water conservation.

8.2 WATER QUALITY PARAMETERS

The quality of water is determined by the presence of substances in the water. To completely describe the quality of water requires the determination of all the substances present in the water and the amount of each substance. Such a determination will be onerous, time consuming and very expensive. Fortunately experience has indicated that it is not necessary, and in many cases the determination of a few simple substances or a combination of substances will suffice. We will introduce the parameters that will help us understand the qualification of water into six grades. We will introduce in a latter part of this chapter other water quality parameters which are relevant to water quality criteria for the major beneficial uses of water. Typical concentrations for domestic waste water are presented in Table 8.1.

The water quality parameters you need to be familiar with to understand the classification of water into various grades are: Suspended Solids, Biochemical Oxygen Demand, Nitrogen, Phosphorus, Pathogens and 'Other pollutants', with Heavy metals being our focus. These terms are explained below.

8.2.1 Suspended solids (SS)

Suspended solids are solids in water that are generally in suspension in the water. Given sufficient time in a settling basin much of the solids will either settle to the bottom or float to the top, and can therefore be removed either as sludge or scraped from the surface as scum. SS is usually measured by filtering a known volume of water and weighing the solids retained on the filter paper after drying.

8.2.2 Biochemical oxygen demand (BOD)

BOD is a measure of the amount of biodegradable organic substances in water. It is thus a parameter that expresses the combined influence of many substances together, their common characteristic being that they are all readily consumed by bacteria in the water. These substances are in fact 'food' for the bacteria. (The scientific term is 'substrate'). In the process of utilising the substrate for growth and energy, respiration takes place and oxygen is consumed. The amount of oxygen consumed is dependent on time and temperature. The higher the temperature the more rapid the uptake of oxygen. BOD measurement is standardised to measure the oxygen uptake over a period of 5 days at 20 °C.

BOD is also a measure of the potential impact of water containing readily biodegradable organic substances if discharged into a water environment. A high BOD water can rapidly deplete dissolved oxygen of the receiving water, rendering it unfit for aquatic organisms which need oxygen for respiration.

8.2.3 Nitrogen (N) and phosphorus (P)

Both nitrogen and phosphorus are plant nutrients. Water containing these elements has fertiliser value. If discharged to a water environment they promote the growth of algae. Excessive growth of algae is undesirable, because when the algae decay they become substrate for bacteria. In the process oxygen is taken up, and at the same time nitrogen and phosphorus in the decaying biomass are released to the water column to fuel the growth of algae. The cycling process continues, algal biomass can accumulate and oxygen may become depleted, particularly in slow or non-moving water bodies (lakes, reservoirs, swamps). The process leading to excessive enrichment of a water body with nutrients is called eutrophication.

8.2.4 Pathogens

Pathogens refer to bacteria, viruses, parasites and helminths that cause illnesses. These are largely derived from human excreta of people who are affected by the pathogens or who are healthy carriers but excrete them nonetheless. You may wish to refer to publications on sanitation, public health or microbiology to gain an appreciation of the life cycles of these pathogens and the role played by water in their transmission through the faecal-oral cycle or through the water environment (e.g. Feacham et al., 1977).

8.2.5 Heavy metals

Heavy metals refer to metals such as Cadmium (Cd), Copper (Cu), Chromium (Cr), Nickel (Ni), Lead (Pb) and Zinc (Zn). These metals affect plant, animal and human lives. Some are essential for plant, animal and human growth (Cu, Zn), but in high amounts can become toxic. Other heavy metals do not play any positive role and become toxic above a certain threshold level, which is specific to each metal and organism.

8.3 WATER QUALITY GRADES

Water quality grades are related to the presence in water of substances described in the above water quality parameters. More importantly the grades are directly related to available water treatment processes. The relationship between the water quality grades, treatment processes and the water quality parameters is shown in Table 8.2.

We begin with used-water, commonly termed wastewater. It is water that has been used for any purpose (domestic, commercial, industrial etc.). Typical values of water quality parameters for water that has been used for domestic purposes are shown in Table 8.1.

Table 8.1. Water quality parameters and typical concentrations for domestic waste water.

Water quality parameters	Typical values for domestic wastewater
BOD	250 mg/L
SS	250 mg/L
N (total nitrogen)	50 mg/L
P (total phosphorus)	15 mg/L
Pathogens	Presence depends on prevalent illnesses
Other pollutants	Dependent on use of chemicals in the household

These are typical values only. When water use per person is high, the concentrations are lower, while concentrations will be higher when water conservation measures are practiced. Use of high-phosphorus detergent will increase P in the wastewater.

When trade wastes are collected with domestic wastewater the concentration of other pollutants can be very high. Heavy metal concentrations from an electroplating plant can be very high unless the trade waste is pre-treated before discharge to the sewer. It is desirable for water conservation purposes to separate waste streams containing high pollutant concentrations and deal with it separately.

Grade 1 water is produced after a simple train of operations involving passing the water through a bar screen, then to a grit chamber and finally through a sedimentation tank. The bar screen, with spacing between bars of the order of 1 cm removes large objects such as napkins. The grit chamber is a sedimentation tank, but the residence time is relatively short, of the order of minutes. Its function is to remove coarse objects that settle relatively quickly, such as sand and similar materials. Aeration may be used in the grit chamber to assist in separating the grit from other suspended solids. In the following sedimentation tank the suspended solids are allowed to settle over a period of the order of hours. The SS of domestic wastewater is usually reduced to about 100 mg/L. The BOD associated with organic materials removed as SS is significant and BOD is reduced to approximately 150 mg/L. There is also removal of nitrogen, phosphorus, pathogens and other pollutants, because some are associated with the SS or adsorbed to the surfaces of the solids. The extent of removal varies and generally as large as BOD removal. Grade 1 water is more usually known as primary treated wastewater, because the treatment process is usually called primary treatment. Waste solids are produced from the bar screen, grit chamber and sedimentation tank, and these need further treatment prior to reuse or disposal.

Table 8.2 Relationship between water quality grades, water treatment and quality parameters

Water quality grades*	Water treatment processes**	Water quality parameters
Used-water (wastewater)		Relatively high values for BOD, SS, N, P, pathogens and other pollutants, especially if it includes trade wastes
Grade 1 (primary treated wastewater)	Primary wastewater treatment (bar screen, grit removal and sedimentation)	Reduction in SS and BOD associated with the removal of SS
Grade 2 (secondary treated wastewater)	Secondary treatment (activated sludge or biological filtration)	Reduction of BOD and SS
Grade 3 (tertiary treated water)	Tertiary treatment (biological nutrient removal)	Reduction in nitrogen and phosphorus
Grade 4	Polishing (treatment processes to achieve quality not dissimilar to water in the open environment)	Reduction in BOD, SS, N, P, pathogens and other pollutants to levels similar to background environmental levels.
Grade 5 (Drinking water)	Treatment of the water to make it conform to the World Health Organization's guidelines for water to be used for drinking purposes	Values of water quality parameters to conform to Water quality criteria for water intended for drinking purposes
Grade 6 (Very high quality water)	Treatment will be specific to meet the intended water quality	Quality depends on the intended use of the water

* Nomenclature for grading is adapted from Simpson and Oliver (1996).

** Treatment processes in this table and accompanying text are described in general terms only. Readers who are interested in these processes should refer to books such as the one by Tebbutt (1992).

Grade 2 water can be produced from Grade 1 water by meeting the oxygen demand of the wastewater. The treatment process relies on bacteria which consume the biodegradable organic substances. Bacteria are retained in the process in two major ways. In the activated sludge process the flocs of bacteria are separated from the treated water, and recycled back to consume organic materials in the incoming wastewater. The recycled bacterial mass is called activated sludge. In trickling filtration the bacteria are retained on the surfaces of a filter medium. The medium can be gravel, but more commonly now plastic sheets on to which the bacteria attach themselves in the form of slimes. Wastewater is distributed evenly above the medium and trickles on its way down the filter. In both processes the organic substances are first adsorbed on the surfaces of the bacterial flocs or slimes and then consumed by the bacteria.

Oxygen is supplied to the bacteria in the activated sludge process by aeration in a tank where the wastewater is contacted with the returned activated sludge. In the trickling filtration process oxygen is supplied through natural convection from below the trickling medium. Sludge is produced in either case, because there is net bacterial biomass produced from consuming the incoming organic materials. The sludge has to be further treated prior to reuse or disposal.

Grade 2 water is better known as secondary treated wastewater, and the treatment process secondary treatment. The BOD of the water by this stage is generally less than 20 mg/L and the SS less than 30 mg/L. The appearance of the water is fairly clear. Nutrients are reduced to about 25 mg/L N and 10 mg/L P. Pathogens are still present in significant numbers.

Grade 3 water can produced from Grade 2 water by removing further nitrogen and phosphorus. The treatment process to achieve the removal of nitrogen is a bacterial process called nitrification and denitrification. The process relies on using nitrifying bacteria to consume ammonia in the water (the predominant form of nitrogen in Grade 2 water) and convert it to nitrate. Readily biodegradable organic substances have become low in concentration by this stage. Only bacteria such as the nitrifying bacteria can derive energy for growth by oxidising ammonia into nitrate. Oxygen for the respiration (oxidation) process is supplied to the bacteria in the same way as in secondary treatment. The nitrate is then provided to another group of bacteria called denitrifying bacteria, which convert it to nitrogen gas. The denitrifying process is intentionally conducted in the absence of oxygen, so that the bacteria have to obtain it from bound oxygen for oxidation purposes, in this case oxygen in the nitrate. Some readily biodegradable organic substances have to be provided to the bacteria for growth. Nitrogen can be reduced readily to less than 10 mg/l and down to 1 mg/L. Further reduction in BOD and SS takes place during the treatment processes, but phosphorus is not affected.

Phosphorus can be removed chemically and biologically. Dosing with alum (aluminium sulphate) or ferric chloride can remove the phosphorus, which is predominantly in the form of phosphate after secondary treatment. A precipitate is produced and will require further treatment for reuse or disposal. Biological phosphorus removal relies on certain bacteria that can take exceptionally large amounts of phosphorus. These bacteria can be cultivated in the process by subjecting them through a cycle of conditions (without oxygen and with oxygen). Phosphorus concentration can be reduced down to 1 mg/L or less. The removal of both nitrogen and phosphorus can be simultaneously achieved with configuration of reactors with and without oxygen.

Grade 4 water is not dissimilar to water that exists in the open environment. Water quality of water environments varies widely ranging from water in pristine lakes containing water replenished from rainfall precipitation to water in swamps which are rich in organic materials, deposits and sediments. We are referring here, however, to water that is of relatively good quality, not different to water in a Class 1 river (Tebbutt, 1992). The BOD of the water does not exceed 3 mg/L, and generally with an average of half of that value. Consequently the dissolved oxygen in the water is above 80% of saturation. The oxygen demand of the water is readily replenished by oxygen transfer from the atmosphere.

Grade 4 water can be derived from grade 3 water by a number of treatment processes. More natural treatment processes include constructed wetlands and sand filtration. In constructed wetlands the wetland plants and the substrate on which they grow help filter the water. Pathogen die-off takes place during flow through the constructed wetlands, which takes place over the order of days or weeks. In slow sand filtration water is passed through a relatively large (area wise) bed of sand at a slow rate. A layer of bacterial slime develops at the top the sand bed that helps further purify the water. In a rapid sand filter a smaller bed of sand is used with a higher rate of water throughput, but the water is dosed with coagulants and/ or flocculants to agglomerate the very small colloidal particles into larger flocs.

Other chemical means of treatment include disinfection using chlorine, ozone or ultra-violet radiation, removal of refractory organic substances using activated carbon, and removal of dissolved salts by membrane filtration.

Drinking water (Grade 5 water) can be produced from Grade 4 water by further treatment using the chemical means described above for producing Grade 4 water. A combination of activated carbon, membrane filtration and ozonation provides a triple-barrier process to ensure that the quality of drinking water is assured.

High quality water (Grade 6) may be required for very specific purposes. For example water for feed into boilers in a power plant should preferably consist only of H_2O molecules. When the water is evaporated any solids will be left behind and scale the boilers. Similarly water used in the manufacture of high quality electronic components should be very pure. Special treatment processes, such as reverse osmosis and ion-exchange in addition to those described to produce drinking water, can be used to produce water of very high purity.

8.4 BENEFICIAL USES OF WATER

The classification of water described above based on available treatment processes is useful, because we have an idea of the quality of water that can be produced using currently available technology. Past experience has also indicated that water of Grade 2 can be used for irrigation of parks and golf courses, provided that the water is chlorinated to reduce the total coliforms bacteria to less than 1000 per 100 mL, and there is several hours of non-access period after the end of an irrigation period. This form of water reuse has been practiced worldwide, particularly in arid areas. Water of Grade 2 has also been used for irrigating crops that are not consumed raw.

It is important, however, to determine scientifically the required water quality criteria for various uses of water. For this purpose a classification of water based on the beneficial uses of the water has been developed. A major classification of beneficial uses is into

- Protection of aquatic systems
- Recreational water
- Raw water for drinking water supply
- Agricultural water use
- Industrial water quality

For each beneficial water use a set of **water quality criteria** have been established. These water quality criteria consist of values for relevant water quality parameters, which should not be exceeded if the water is to be used for the intended beneficial use. These water quality criteria have been derived from scientific investigations into the effects of pollutants on plants, animals, humans and the environment. The best known set of water quality criteria is one for drinking water (WHO, 1993, NHMRC, 1996). As our knowledge of the effects of pollutants increases we need to update the water quality criteria. Australia has developed water quality criteria for all the beneficial uses of water mentioned above (ANZECC, 1999).

8.5 PRACTICAL CONSIDERATIONS

8.5.1 Critical parameters

In practice it is not feasible to determine the presence and concentration of all listed parameters of a particular water when determining whether it is suitable for an intended use. The cost of water analysis will be prohibitive and the time required may be considerable. It is unlikely that all the water quality parameters in a set of water quality criteria will be significant (compared to the threshold value that should not be exceeded). Two considerations will help in determining which parameters should be measured. One is the principle of conservation of matter. In a system any material entering it (raw materials, including input water, chemicals added) will leave in the products or waste streams. Knowing the quality of the input streams will assist in determining the substances in water leaving the system. The second is that it is likely that similar systems have been studied previously. Searching the literature for references to similar systems is therefore recommended.

8.5.2 Sampling and analysis of water

Analysis of water for the presence and concentration of specified substances is best done by a laboratory that is certified to carry out such analyses. It is important to enquire about the method used for analysis, or better still to specify the method to be used. A useful reference is the 'Standard Methods for the Examination of Water and Wastewater' published by the American Public Health Association, American Waterworks Association and Water Environment Federation (1995). Such a laboratory will also specify the volume of sample required, provide the necessary sample container and method of storing and transport (e.g. whether acidification or refrigeration is required). Samples for determination of trace quantities of heavy metals, for example, should be collected in containers whose surfaces do not adsorb heavy metals. It is desirable to rinse the container with the sample to be taken. Samples for determination of pathogens should be collected in containers which have been sterilised. Collection of the samples should also follow techniques which prevent the samples from being contaminated by coming in touch with human hands. The samples should then be stored in ice and delivered to the laboratory as soon as is feasible and preferably within 24 hours. Consultation with a certified laboratory and following the direction given is generally good practice.

Many water streams do not have a constant flowrate or water quality. It is important to obtain a representative sample of any water stream. This can be done by sampling over a time period and compositing the samples. Weighting based on flowrate is a preferred way of obtaining a representative sample. In many cases this may not be feasible, and grab samples may have to do. Results should be interpreted carefully. A useful reference on obtaining representative samples and conducting checks on results to ensure that the principle of conservation of matter is satisfied is Eckenfelder (1980). The composition of a wastewater stream generally varies with time of day and over the week. The latter may be due to differing operations being carried out during the week. Statistical techniques are available to determine the average value over a time period as well as values that will not be exceeded 90% or 95% of the time.

8.6 REFERENCES

APHA, AWWA, WEF (1995). Standard Methods for the Examination of Water and Wastewater, Jointly published by the American Public Health Association, American Water Works Association and Water Environment Federation, Washington, DC.

Australia and New Zealand Environment and Conservation Council (1999). Australian and New Zealand guidelines for fresh and marine water quality [computer file] : draft / prepared under the auspices of Australia and New Zealand Environment and Conservation Council and Agriculture and Resource Management Council of Australia and New Zealand

Eckenfelder, W.W. Jr. (1980). Principles of Water Quality Management, CBI Publishing, Boston.

Feacham, R, McGarry, M and Mara, D.D (Eds.). (1977) Water, wastes and health in hot climates, Wiley-Interscience, London.

National Health and Medical Research Council (1996). Australian Drinking Water Guidelines, NHMRC, Canberra.

Simpson, J. and Oliver, P. (1996). Water quality. From wastewater to drinking water to even better. Australian Water and Wastewater Association, Sydney.

Tebbutt, T.H.Y. (1992). Principles of Water Quality Control. Pergamon, Oxford.

World Health Organization (1993) Guidelines for drinking water quality, World Health Organization, Geneva.

9

Environmental legislation – impact on water

9.1 INTRODUCTION

We address here the second of the two outstanding matters from Chapters 3 through to 5, in which we have worked through the 'Water audit process' as visualised in Figure 2.2. This chapter provides an overview of environmental protection legislation, particularly as it relates to the management of water related issues including water auditing. The major forms of environmental regulation are described including environmental impact assessment and pollution prevention and control. We also discuss recent developments in non-regulatory approaches.

 ISBN: 1 900222 52 3. Published by IWA Publishing, London, UK.

9.2 ENVIRONMENTAL PROTECTION

9.2.1 Legislative approaches

Environmental protection regulation has existed in various forms at least since Roman times (Thomas, 1998) and early forms of regulation of water related issues focused primarily on the protection of public health. For example, Australia's first pollution regulations aimed to prevent felling of trees within 15 m of the Tank Stream, which was Sydney's only water supply for the first five years of settlement. This regulation was later strengthened by proclamation to prevent "throwing of any filth into the Stream of fresh water...". Until the late 1960s, most environmental regulation was based around Health Acts and statutes relating to control of drinking water catchments. This form of regulation has continued and is now generally known as pollution control or pollution prevention and concentrates on the management of existing activities to prevent or minimise environmental harm or risk to human health.

Modern environmental protection legislation can be traced to the US *National Environmental Policy Act 1969*. This legislation formally introduced a new concept of environmental control or management known as environmental impact assessment (EIA) which has since been adopted in various forms around the world (Harvey, 1998). EIA is intended to assess the potential environmental consequences or impacts of proposed activities and to identify options to minimise environmental damage. In general usage, EIA is used to inform decision makers of the likely environmental impact of a proposal before a decision is made, and to provide an opportunity to identify key issues and stakeholders early in the life of a proposal (Environment Australia, 1999).

Pollution control legislation initially tended to concentrate on specific media, typical examples being the United States *Clean Air Act 1977*, the *Clean Water Act 1977* and other legislation dealing with issues such as pesticides and toxic substances (Miller, 1994). Many early examples of environmental permits were consequently focused on specific environmental media. This style of regulation was often driven by the fragmented nature of regulatory responsibility. For example, water authorities tended to focus on water quality issues while health departments focusing on waste disposal, odours and sometimes on contaminated sites.

More recently, with the creation of specific environmental protection regulatory bodies, there has been a general transfer and concentration of responsibilities and a consequent move towards integrated pollution control regulation. This has led to environmental assessment procedures, licensing systems which cover emissions to all environmental media - land, air and water, as well as other areas of environmental monitoring, policy development, land-use planning and ecological protection.

The approach taken in Australia to environmental legislation has tended to follow trends in the United States and all Australian states and territories enacted some form of environmental protection legislation during the 1970s and 1980s. This legislation generally provided for both pollution control and EIA processes, but did not provide comprehensive coverage of environmental protection issues, with various powers remaining vested in local government authorities, planning authorities, health departments and water supply utilities. A more detailed summary of environmental law in Australia is provided by Bates (2002).

The Commonwealth of Australia government has also enacted federal level environmental protection legislation, although this has concentrated on EIA of projects involving Commonwealth lands or entities or involving international treaty obligations. The major Commonwealth legislation is the *Environmental Protection and Biodiversity Act 1999* which deals with issues relevant to international treaty obligations, Federal government bodies (e.g. defence, airports) and proposals of national significance.

Increasingly, application of environmental regulation is being based on a common set of ecologically sustainable development concepts including:

- the polluter-pays principle;
- requirements for intergenerational equity;
- conservation of biological diversity; and
- the precautionary principle.

These principles have been written into legislation or used as the basis of environmental policy-making at both government and industry levels.

9.2.1 Non-regulatory approaches

There have been a number of recent developments in the field of non-regulatory environmental management. The International Organisation for Standardisation (ISO) has developed a series of environmental management systems standards known collectively as the ISO14000 series, which formed the basis for our discussion in Chapter 2. These systems apply similar principles to the earlier quality management systems (ISO 9000 series) and encourage companies to adopt structured and documented environmental management principles. Water audits may be required in the processes of determining aspects of environmental management systems, in determining appropriate remedial actions to environmental issues, and in monitoring and measurement of performance.

Other initiatives such as cleaner production, environmental improvement programs, and voluntary self-regulation through industry codes of practice have provided opportunities for the management of environmental outcomes outside of a legal framework. The International Chamber of Commerce and the World

Business Organisation have developed a Business Charter for Sustainable Development which actively promotes the cleaner production philosophy (ICC/WBO, 1999), while the Plastics and Chemical Industries Association has developed the Responsible Care Program (PACIA, 1999, Gunningham et al., 1998). Similarly the International Council for Local Environmental Initiatives, an international movement of local governments and their stakeholders, has embarked upon 'The Water Campaign' (Morrison & Brugmann, 2000). This campaign seeks to work with both higher levels of government and community members to promote and achieve use of fresh water in a more sustainable way in their locale.

9.2.3 Environmental impact assessment

According to the International Association for Impact Assessment (IAIA) The objectives of environmental impact assessment (EIA) are (IAIA, 1999):

- to ensure that environmental considerations are explicitly addressed and incorporated into the development decision making process;
- to anticipate and avoid, minimise or offset the adverse significant biophysical, social and other relevant effects of development proposals;
- to protect the productivity and capacity of natural systems and the ecological processes which maintain their functions; and
- to promote development that is sustainable and optimises resource use and management opportunities.

Furthermore, EIA should be purposeful, rigorous, practical, relevant, cost-effective, efficient, focused, participatory, interdisciplinary, credible, integrated, transparent and systematic. In addressing these basic principles, the EIA process should provide for screening, scoping, examination of alternatives, impact analysis, mitigation and impact management, evaluation of significance, preparation of an environmental impact report, review of the report, decision making, and follow-up.

These elements will be outlined for the example given below, the Western Australian EIA process.

9.2.4 Environmental licensing

Environmental licensing is closely associated with the concepts of pollution control and enforcement. Licences or permits are issued to regulate emissions from activities that are considered to be potentially polluting. Terms such as prescribed activities, prescribed premises, and noxious industries are commonly used to define industries that are licensed.

The principal aim of an environmental licensing process is to limit discharges to levels that ensure protection of the receiving environment, or to impose a regulatory compulsion to limit or reduce emissions according to some timeframe. There have been a number of different approaches to licensing and these are discussed briefly below.

During the 1970s and 1980s, the USEPA adopted what has become known as a "command and control" approach to regulation in which environmental laws were highly prescriptive as to what could or could not done (Comino and Leadbeter, 1998, Gunningham et al., 1998). Industry was expected to adopt uniform best available technology pollution controls as specified by the regulatory authorities and non-compliance was met with strict enforcement through the legal system. This litigious approach was confrontational and led to considerable diversion of industry and regulatory resources away from environmental protection to legal disputation. Some regulatory agencies in countries outside the USA adopted similar styles of operation, but generally were less able to match the necessary resource requirements.

In other countries, including Australia, and more recently in the USA, there has been a move towards an approach that places greater emphasis on environmental outcomes as the basis for pollution control. This approach is founded on the principles of pollution prevention, integrated pollution control and optimising the regulatory mix. Using this approach, pollutant emission controls are applied at a level sufficient to achieve the required receiving environment quality. In some instances this will require best practice controls, and in others less stringent controls. More importantly, the regulator no longer specifies how emissions are to be controlled, but sets the required environmental performance. Coupled with this, the imposition of enforcement measures is used to encourage compliance rather than punish non-compliance. The role of the regulator has become less of an enforcer of standards and more of a partner and facilitator.

To encourage further voluntary compliance with environmental laws, or even better performance, a number of incentive schemes have been developed. The USEPA has been running a re-invention program since 1995, which includes a wide range of initiatives intended to "motivate superior environmental performance" (USEPA, 1999). This program includes such elements as encouraging development of environmental management systems and self-auditing as well as reduced levels of regulatory burden.

In other areas government funded cleaner production programs have been established to provide financial and technical assistance to improve environmental performance. Two examples are the programs operated by the Environmental Protection Group of Environment Australia (Environment Australia, 1999) and the Victorian EPA (Vic EPA, 1998).

More direct incentive schemes include alternative forms of licences, which apply different levels of control, and fees, to industries depending on the degree of environmental risk that they pose. Various forms of best practice licence have been developed which provide incentives in terms of reductions in the degree of regulatory attention and fees. In general this type of licence applies the concept of audited self-management rather than self-regulation.

Another recent innovation is the use of Environmental Improvement Plans (EIPs). EIP systems have been implemented in both South Australia (South Australian EPA, 2003) and Victoria (Victorian EPA, 2003) and have the general objective of delivering improved environmental performance over time under an agreed framework developed cooperatively by industry, government and the community. A similar approach has been adopted to manage Water Corporation wastewater treatment facilities in Western Australia, and a more general EIP system is being developed for other regulated industries.

9.3 CASE STUDY: WESTERN AUSTRALIAN ENVIRONMENTAL LEGISLATION

9.3.1 Background

The Case Study uses the regulatory system in Western Australia as a typical example of an environmental control system and also provides some specific case studies to illustrate the application of water auditing principles that have arisen from the application of regulatory requirements. The sovereign state of Western Australia forms part of the federation of similar states forming the Commonwealth of Australia. The government of Australia as a whole is called the Federal or the Commonwealth government in this discussion. The Australian governmental system derives from the Westminster system, thus having similarities with other federal Commonwealth countries such as Canada. Yet Australia also has governmental features in common with those of the United States of America. Australia has a three-tiered system of government (unlike Great Britain) of Federal, State and Local governments. It thus encompasses diversity and complexity, which makes it a good example for uncovering issues of law that are of interest to the water auditor in a wide range of other countries, some of which will have less complex legal arrangements. Though the example used will not be precisely applicable in other countries, the frameworks, and most issues to attend to, will be flagged by the case study of Western Australian regulation.

In Western Australia, water related environmental issues are dealt with in five major pieces of state legislation. These are:

- The *Environmental Protection Act 1986*
- The *Water and Rivers Commission Act 1995*;

- The *Rights in Water and Irrigation Act 1914*;
- *Metropolitan Water Supply, Sewerage, and Drainage Act 1909* and associated by-laws; and
- The *Health Act 1911*.

The following discussions focus primarily on the operations of the *Environmental Protection Act 1986* (the Act) as a typical example of environmental protection legislation. The Metropolitan Water Supply by-laws are also described as they are a useful example of direct industry discharge controls.

The *Rights in Water and Irrigation Act 1914* will not be discussed in detail, but it does include provision for restricting the amount or rate of water that can be taken from a water resource. Information about water use is available through a publicly accessible register of licences. This licensing process may indirectly stimulate water auditing as a means of establishing a legitimate water need or to identify opportunities for water use reductions in order to meet licensed water abstraction limits.

The Act places requirements on operators of industrial premises or proponents of development to meet certain environmental objectives and provides for the establishment of the Environmental Protection Authority (EPA) and defines the respective powers of the EPA and the Department of Environmental Protection (DEP).

The Act provides the legislative basis for management of all environmental issues in WA including water issues, and relates to water management aspects principally through Part IV (Environmental Impact Assessment) and Part V (Control of Pollution). However, Part III of the Act provides for development of environmental protection policies that could, as is the case in Victoria, deal with issues relating to setting of water quality objectives and preservation of water quality.

9.3.2 Environmental impact assessment (EIA)

Part IV of the Act, which deals with EIA, has the following features:

- it is implemented by the EPA;
- the DEP provides advice and administrative assistance;
- it provides for assessment of proposals;
- it leads to the setting of legally binding Ministerial Conditions.

In its application to water auditing, the EIA process may be applied to issues related to quantity, quality or a combination of these. In relation to water quantity, a water balance is usually required as part of the overall industrial or mining process design as it is critical to determine the feasibility of some processes in a water-deficient environment such as is the case in much of WA. To date this has not generally been an explicit part of the EIA process, but no doubt water auditing does occur as a support activity to the preparation of EIA documentation.

Water quality related drivers for the audit may include raw water quality limitations or wastewater quality objectives, the latter being driven by proposed wastewater discharge limits or reuse requirements.

The aim of the EIA process in relation to water issues is to ensure the sustainability of water usage and the environmental acceptability of any proposed discharges. Depending on whether a proposal is for an entirely new project or expansion of an existing facility, water auditing activities associated with the EIA may be theoretical and/or based on operating experience.

The EIA process in Western Australia is outlined as a flow chart in Figure 9.1. Through a public process, the independent EPA undertakes the EIA process in WA. The key elements of the process for formal assessment of a project are:

- the proponent is required to prepare a public document which outlines the proposal and its environmental setting;
- this document, usually called a Public Environmental Review (PER) or Environmental Review and Management Program (ERMP), is then released to the public for a comment period ranging from 4 weeks to 10 weeks;
- submissions are summarised and the proponent is required to prepare a written response;
- the EPA considers all of the information, comprising the proponent's document, submissions, response and any other information, and prepares a report of advice to the Minister for the Environment;
- this EPA report, including the proponent's response to submissions, is released for public appeal by the Minister;
- following determination of appeals, the Minister decides whether, and under what legally binding conditions, the project should be implemented.

9.3.3 Control of pollution

Part V of the Act has the following features:

- it is implemented by the Environmental Regulation and Regional Operations Divisions of DEP;
- instruments available include registration, works approvals, licences, pollution abatement notices and directions;
- legally enforceable regulations apply to registrations and other aspects (e.g. noise, air quality, ozone depleting substances);

- legally binding conditions apply to works approvals and licences.

Part V of the Act provides for the licensing of prescribed premises. DEP licenses and works approvals contain conditions aimed at regulating discharges to the environment. DEP licenses have always been comprehensive environmental licenses, covering air, water and other waste emissions.

Licence and works approval conditions must conform to the requirements of section 62 of the Act and may require:

- installation and operation of specified pollution control equipment within specified times;
- taking measures to minimise the likelihood of pollution arising;
- provision of monitoring equipment of a specified type;
- monitoring relating to characteristics, volume and effects of discharges or emissions;
- reuse of waste or the making of waste available for reuse by another person; and
- operation of equipment to prevent pollution.

All licences and works approvals contain conditions, which are based on certain generic licence condition sets. However, the conditions are always tailored so that they are specific and relevant to the individual premises.

Regulations which support registered premises follow similar lines to licence conditions but apply equally to all premises within a given registration category.

The licensing system in Western Australia aims to target regulatory effort on those premises with the greatest potential to pollute. In line with this objective, prescribed premises are classified into three groups:

- Schedule 1 Part 1 - Category 1 to 67A Works Approval and Licence
- Schedule 1 Part 2 - Category 68 to 87 Works Approval and Registration
- Schedule 2 Registration only

Some examples of prescribed premises definitions in Schedule 1 of the regulations are given in Table 9.1.

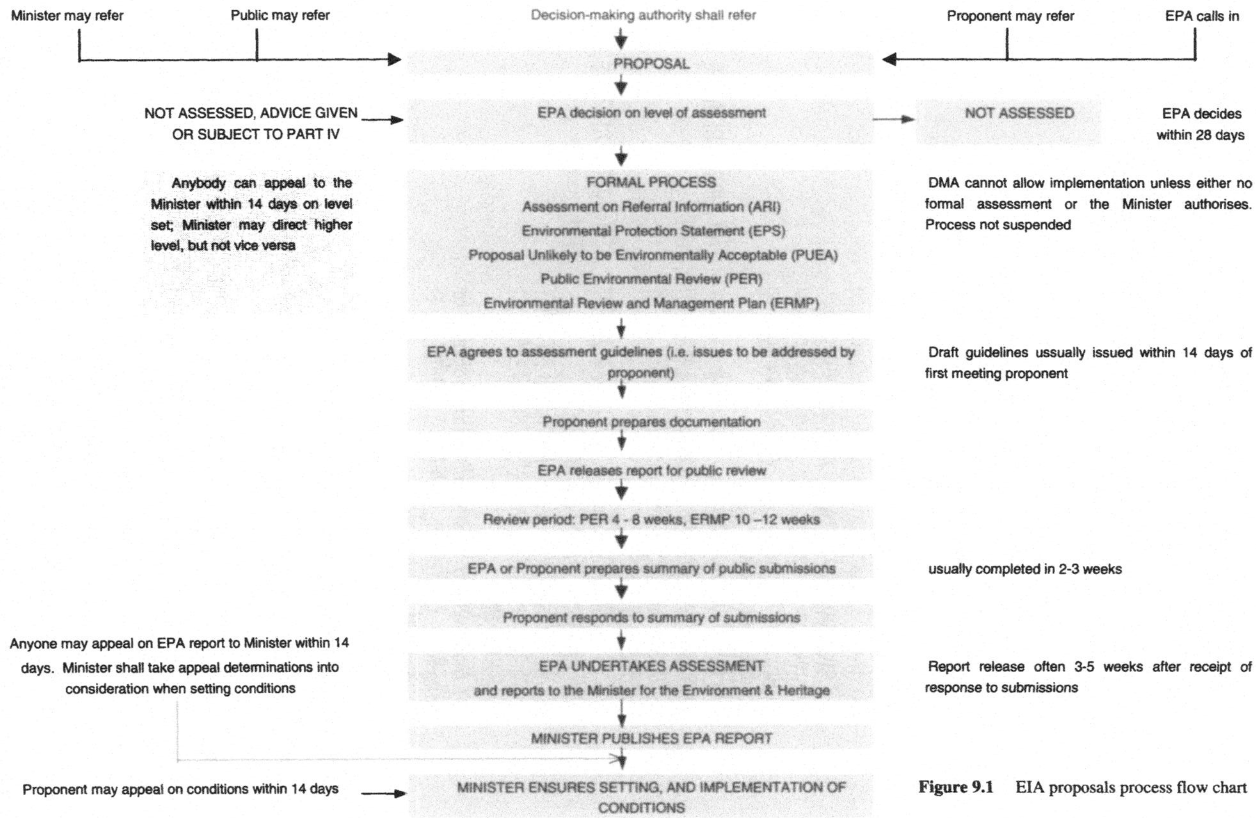

Figure 9.1 EIA proposals process flow chart

Figure 9.1 EIA process in Western Australia.

Table 9.1 Examples of prescribed premises definitions

Category number	Description of Category	Production or design capacity
3	Aquaculture (ponds or tanks): premises on which - (a) marine, estuarine or freshwater fish or prawns are propagated or reared; and (b) supplementary feeding occurs, in ponds or tanks that discharge waste into waters or onto land.	Biomass 1 000 kilograms or more
6	Mine dewatering, tailings or residue disposal: premises on which - (a) water is extracted and discharged into the environment to allow mining of ore; or (b) mining or processing of ore occurs and tailings or residue are discharged into a containment cell or dam.	50 000 tonnes or more per year
31	Chemical manufacturing: premises (other than premises within category 32) on which chemical products are manufactured by a chemical process.	100 tonnes or more per year
61	Waste treatment facility: premises (other than premises within category 54) - (a) on which liquid waste produced on other premises is stored, reprocessed, treated or irrigated; or (b) from which waste, including septage or sewage treatment plant sludge, is discharged onto land.	100 tonnes or more per year

9.3.4 Licence fee structure

The works approval, licence and registration fee structure was set up as follows:

- Works approval: Sliding scale based on the capital cost of works
- Registration: Once-only application fee of $300
- Licence: Fee based on:
  - (a) Category of prescribed premises
  - (b) Production or design capacity
  - (c) Discharges to the environment

The discharge fee is based either on assumed or monitored emissions and is waived for Best Practice Licenses. Table 9.2 gives some examples of how fee units are assigned to different kinds of discharges.

Table 9.2 Example of discharge fees (waste onto land or waters)

No	Kind of waste	Fee units
1	Liquid waste that can potentially deprive receiving waters of oxygen (for each kilogram discharged per day) -	
	(a) biochemical oxygen demand (in the absence of chemical oxygen demand limit)	0.5
	(b) chemical oxygen demand (in the absence of total organic carbon limit)	1.0
	(c) total organic carbon	0.5
2	Biostimulants (for each kilogram discharged per day) -	
	(a) phosphorus - (i) Swan Coastal Plain	10
	(ii) elsewhere	2
	(b) total nitrogen - (i) Swan Coastal Plain	10
	(ii) elsewhere	2

9.3.5 Best Practice Environmental Licences

Best Practice Environmental Licenses (BPELs) are a new form of licence class and are based on the principle of audited self-management. The essential elements of a BPEL are:

- environmental performance is benchmarked to best practice and agreed with DEP;
- operation of an Environmental Management system, with third party or DEP accreditation, to ensure continual improvement;
- independent auditing of environmental performance;
- public reporting of performance and pollution incidents.

The criteria for acceptance as a Best Practice Licensee are:

- an environmental policy;
- environmental performance objectives agreed with DEP;
- environmental management plan;
- preparation of implementation program;
- implementation of monitoring program;
- benchmarking environmental performance;
- community involvement;
- reporting of environmental performance; and
- acceptable audit plan.

9.3.6 The need for water management and audits

Improved water management through the conduct of water audits may be driven by the requirements of the DEP licensing system. In particular, for monitored and Best practice environmental licences:

- additional information may be required by DEP;
- both require current and verified waste audit and mass balance calculations;
- water audits are of use to the licensee in assessing high volume water uses or high concentration waste streams;
- environmental management plans required for BPELs will often require water audits;
- continuous improvement plans will also require periodic water audits.

The DEP also has the power to require water audits or improve water management through licence condition setting (Section 62 of the Act). The condition setting powers

- allow the DEP to specify the type of water management equipment to be used to control discharges;
- may require the "reuse of waste wholly or in part...";
- may require the licensee to make the waste available for reuse by someone else.

9.3.7 Specific Case studies

9.3.7.1 Cement works and lime-sand dredging

The background to this case study is as follows:

- shellsand is dredged from the ocean, washed, and pumped approximately 3 km as a slurry from the on-shore washing plant to the cement works;
- brackish water from the stockpile infiltrates the groundwater;
- recovery bores abstract the saline groundwater, which is pumped back to the washing plant for washing of sea sand;
- DEP licence conditions require a net export of salt from the groundwater at the plant such that:
  - (a) the quantity of salts exported from the works during any 90 day period, is equal to or exceeds the quantity of salts imported via the sea-sand slurry pipeline, during the same period.
  - (b) the quantity of recovered groundwater exported from the site is less than 120 percent of the water imported via the sea-sand slurry, during any 90 day period.
  - (c) the ambient salinities of groundwater in various strata beneath the licensee's site are not substantially increased as a result of the importation of saline slurry waters.

Net monthly movement of salt from the works is calculated as follows:
$(TDS_{import} \times V_{import} - TDS_{export} \times V_{export}) / 1 \times 10^6$

where: TDS = total dissolved solids (mg/L)
V = volume of water (m^3)

Some example data is shown in Tables 9.3 and 9.4. These data show that during the period July to September the licence conditions were met. There was both a net export of water and salt to the ocean over the three month period, even though in August there was a net deficit of salt export.

Table 9.3 Water balance July - September

Month	Export water (m^3)	Import water (m^3)	Export: Import
July	93,674	84,570	1.11
August	103,237	94,279	1.10
September	113,952	114,103	1.00
Total	310,863	292,952	

Table 9.4 Salt balance July – September

Month	Salt import (t)	Salt export (t)	Net salt movement
July	206.1	215.7	9.6
August	242.6	231.0	-11.6
September	273.5	279.6	6.1
Total	722.2	726.3	4.1

9.3.7.2 Chemical plant

The background for this case study is as follows:

- The plant produces polyacrylamide flocculants and dispersants for mineral processing and effluent treatment;
- Wastewater streams consist of polymeric water based solutions of polyelectrolytes and monomers;
- Most wastewater comes from vessel washing and general operational cleaning systems;
- Wastewater was previously directed to an unlined soakage pit at the rear of the works;
- Analysis of the wastewater showed very high TDS (~8,000 ppm), and bore water samples showed a significant impact on groundwater quality.

Water use identified in water audit was as shown in Table 9.5. A final water flow diagram is shown in Figure 9.2.

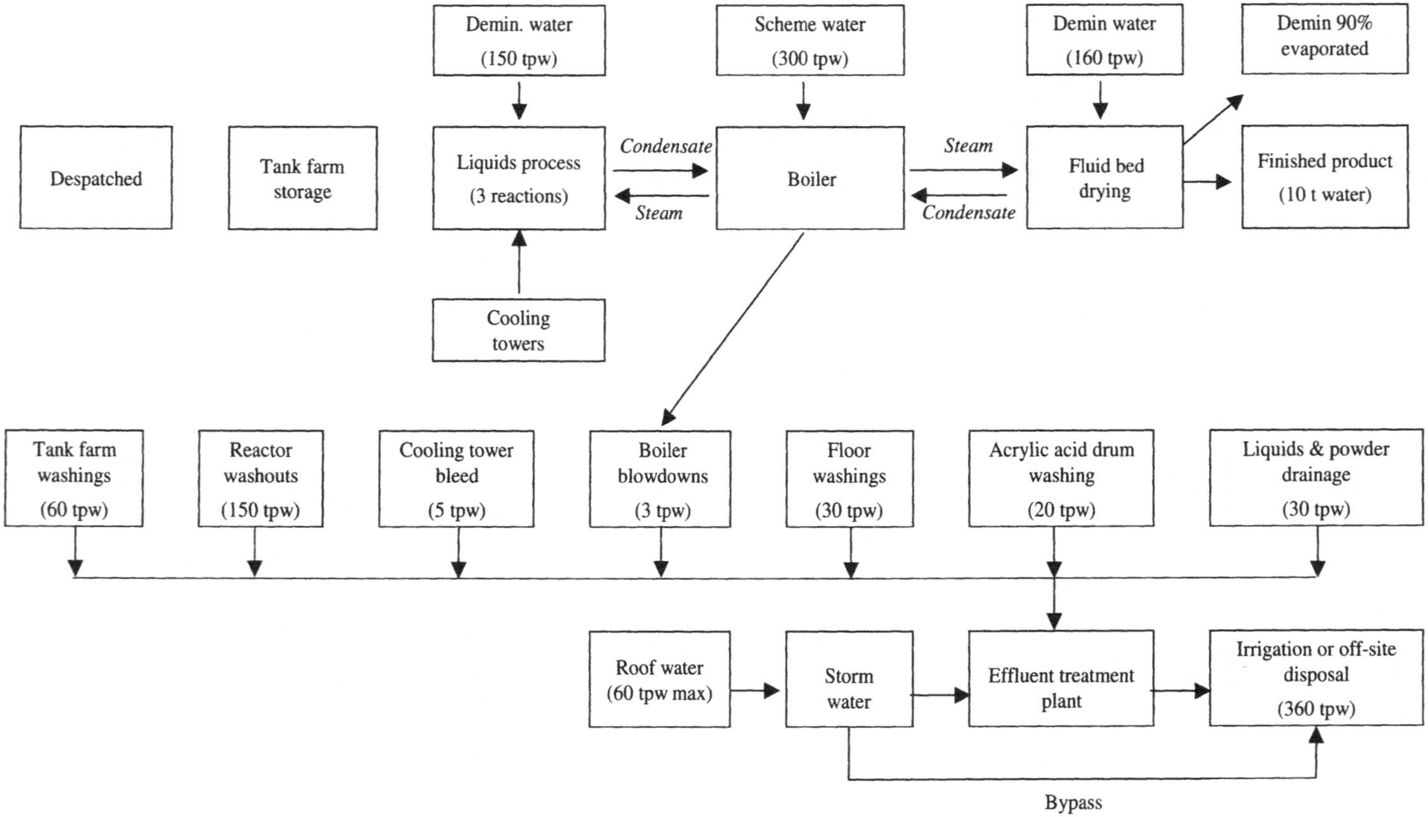

Figure 9.2. Final water flow diagram

Table 9.5 Preliminary water audit findings

Source	tonnes/week	BOD (mg/L)
Reactor effluent (three reactors)	150	
Liquids plant drain	50	1100
Powder plant drain	50	680
Tank farm bunds	60	60
Storm water drains	60 (max)	35
Acrylic acid drum cleaning	20	
Cooling tower bleed	5	
Emulsion reactor washwater	?	<5

Major sources of high BOD wastewater were identified as:

- Spillage of finished product (1 kg of product needs 200 L of dilution water to remove);
- Powder plant polymerisation section drain;

Proposed solutions (the water management strategy, see Chapter 4) for reducing the load of salt and BOD to soakage to acceptable levels were divided into two groups. Short term actions included:

- installing collection tanks for waste streams (~60 m^3 per week now sent off-site);
- installing hard piping (avoiding use of flexible hoses) to reduce general water use;
- purchasing a high-pressure cleaning unit to reduce water use.

Medium term actions included:

- removing contaminated soil to reduce leaching potential;
- installing an effluent treatment plant;
- completing hard piping of liquid process storage areas;
- installing wash-water points with jet nozzles.

9.4 REFERENCES

Bates, G. 2002. *Environmental Law in Australia* 5th Edition. Butterworths, Australia.

Comino, M, Leadbeter, P. 1998. Enforcement of Pollution Laws in Australia - Past Experience and Current Trends. *Fifth International Conference on Environmental Compliance and Enforcement*. November 16-20, 1998. Monterey, California, United States.

Environment Australia. 1999. www.environment.gov.au/net/eianet.html.

Gunningham, N, Grabowsky, P, Sinclair, D. 1998. *Smart Regulation. Designing Environmental Policy*. Clarendon Press, Oxford.

Harvey, N. 1998. *Environmental Impact Assessment*. Oxford University Press, Australia.

International Association for Impact Assessment. 1999. www.iaia.ext.nodak.edu/IAIA.html.

ICC/WBO. 1999. www.iccwbo.org/home/menu_environment.asp.

Miller, A. S. 1994. In *Environmental Outlook. Law and policy*. Boer, B, Fowler, R, and Gunningham, N (Eds). The Federation Press, Sydney.

Morrison, K & Brugmann, J. 2000. *The Water Campaign*. International Council for Local Environmental Initiatives World Secretariat, Toronto, Ontario.

PACIA. 1999. www.pacia.org.au/environment2.html.

South Australian EPA, 2003. www.epa.sa.gov.au.

Thomas, I. 1998. *Environmental Impact Assessment in Australia. Theory and Practice*. The Federation Press, Sydney.

USEPA. 1999. www.epa.gov/reinvent.html.

Victorian EPA. 1998. *Industrial Waste Strategy Zeroing in on Waste. Pathways to Cleaner Production for Victorian Industries*. Environmental Protection Authority, State Government of Victoria.

Victorian EPA, 2003. www.epa.vic.gov.au.

10

Arena: Commercial water use inside buildings

10.1 INTRODUCTION

With the completion of Chapter 9 we have covered the context, principles, strategies and major tools for water conservation and water auditing as exercised in the complete water auditing process. We now change the emphasis, with a series of chapters that examine particular issues, which might confront the water auditor in a variety of typical arenas. An arena is a general setting within which a water audit might be conducted. This chapter will include a small case study. (In Chapter 23 two case studies will be presented one in the form of a water audit report.) Thus we come to the first of the 'arena' chapters, commercial water use inside buildings. The word 'commercial' ought not to be interpreted too narrowly. Domestic residences (for which Water and Energy Savers (1994) is a useful work), office complexes, hotels, motels, hospitals, shopping centres, universities and even accommodation and catering on mining sites are examples falling within the scope of this chapter. Manufacturing and associated processes, mining sites *per se*, landscaping, external commercial use, aquatic centres, sporting and recreational complexes and remote communities, among others, will receive attention in later chapters.

10.2 ACTIONS FOR THE WATER AUDITOR

The actions identified here for the water auditor to undertake are to be read within the context of the previous chapters, and do not supersede them, but rather identify particular matters needing attention within the earlier framework.

10.2.1 Determining water sources and flow rates

Initially, the water auditor identifies the sources of water: scheme-water, bore water, rainwater and, perhaps, other sources too. In many instances scheme water will be the sole source for the building. Whatever sources are identified, the type of flow-meter installed should be noted. Where there is no flow meter installed, the water auditor selects the means of measuring the flow. In the typical situation where scheme water is the only supply source, it will usually have one (or more) permanent, positive displacement, cumulative, in-line water meters. Meter readings can be taken over the period of the audit to gain an exact measurement of the input flow to the commercial premises. When the water audit results are to be interpreted on an annual basis, the water bills from the water service provider are invaluable in providing accurate water usage records over the annual cycle. They might also provide breakdowns of the water usage over smaller periods of time than a year. Concurrently energy costs are determined from gas and electricity providers' bills, to enable later financial evaluations to be undertaken. Reduction of energy usage by means of reduction of hot water use or minimising hot water entropy increase is a part of the normal activity of a water auditor in the commercial arena. While water auditing and energy auditing are not identical, the water auditor Historical records of water, electricity and fuel usage can be suggestive of possible future trends.

10.2.2 Gathering site and occupancy information

In order to obtain a reasonably accurate estimation of the number of water outlets and their purpose, the auditor seeks computer records or hard copy drawings of the building(s)' plumbing system. Together with this information the occupancy rates of the building, especially seasonal variations, are gathered. The auditor can then calculate water usage on the basis of published tables of per capita usage for the geographical area in question. We will address this issue below.

10.2.3 Determining water usage and flow rates

Per capita information on water usage for appliances/outlets may not be available for a specific location. In such circumstances data sheets or questionnaires can be distributed to appropriate personnel seeking information on water usage. We can obtain the most accurate data by measuring individual taps and hand-basins (for example) using a measuring container and stop-watch. In large buildings with many similar water appliances/outlets, a statistical sample may be taken, rather than measuring absolutely every individual appliance/outlet. (Such a procedure lessens the opportunity for detecting leaks.) Typical 'appliances' for which water flow measurements are required are: toilets, showers, hand-basins, urinals, kitchens, cleaners troughs, outdoor taps attached to the building, drinking water fountains and coolers, evaporative air conditioners (evaporation and bleed), cooling tower makeup water when refrigerated air conditioning is installed, boiler make-up water. Water might be used for additional cooling purposes within a commercial building, eg. cooling water for laboratory equipment. Strictly, wastewater discharges should also be measured to attain a complete water audit. Bills from the sewer service provider might provide the required information. Unless there are special reasons for measuring wastewater, it is customary for water auditors to neglect this portion of the rigorous audit process in commercial premises.

10.2.4 Published rates of water use

Water usage rates depend upon climatic and geographical factors, pricing structures and cultural considerations. Thus here we can give only indicative rates. It is essential to obtain data that is correct for a particular city or water service provider's area when conducting an audit. Our indicative figures derive from a medium sized city, located in a warm Mediterranean climate, within a developed country with high quality building and services and a growing culture of careful water use.

Table 10.1. Indicative water usage rates of 'old' and water efficient appliances. Source: Wajon (2000), with modifications.

'Appliance'	Water consumed per use (L)	Water consumed per use with water efficient 'appliances' (L)
Toilet	(Old single flush) 11	(Dual flush 9/4.5 L) 7 (Dual flush 6/3 L) 4.5
Urinal	12	3
Hand-basin	2	1.2
Shower	63	38
Washing machine	120	78
Laundry trough	10	6
Kitchen sink	9	5.5
Drinking fountains	0.2	0.05

10.3 OPTIONS FOR PROMOTING WATER CONSERVATION

Two main ways to promote water conservation exist within the arena of commercial internal water use. The first way involves changing the characteristics of the whole water circuit in a building or series of buildings by means of devices such as pressure reducing valves or flow reduction devices. The second way is to adjust the water usage characteristics of appliances themselves. In some instances there is overlap of the two ways.

10.3.1 Pressure reducing valves and pressure balancing a system.

Scheme water supplied by a water service provider will sometimes fluctuate in pressure on the one hand and sometimes be supplied at a much higher pressure than required on the other. Installing a pressure reducing valve on the user's side of the water meter will save water, while minimising pressure fluctuations and thus variable water use by outlets and appliances.

The aim of pressure balancing a system is to have an equal water pressure at the point of delivery of water to every 'appliance' in the system. Thus, for example, the hot and cold water taps servicing a shower would have the same water pressure and, further more, turning on either of them would not cause significant pressure changes at other points in the system. The outcome of this is that the shower water can be set at a constant temperature; the temperature will not fluctuate with time. The user of the shower is therefore assured of a comfortable shower at the temperature and flow rate of choice. More importantly the shower will be safer to use, without the danger of scalding due to unwanted temperature fluctuations.

To pressure balance a system is equivalent to ensuring that the resistances to water flow in the distribution pipe system are considerably less than those at the actual delivery points for all water flows up to the maximum flow rate deliverable at every outlet. This outcome can be partially achieved by reducing the pressure loss in distribution pipes resulting from reducing the maximum flow rate through them. As discussed above, a pressure reduction valve at the delivery point of water to the building (say) can effect a reduction in pipe pressure losses. As pressure losses in pipes are proportional to the square of the velocity of water through the pipes, modest pressure reductions at the point of entry to the building can have a major impact on pipe pressure losses. What is more, the procedure will even out the pressure at points of delivery due to the non-linear relationship between water velocities in pipes and the associated pressure losses. Increasing the resistances to water flow at the outlets will further aid in the balancing of the system.

The method of doing this is often the use of flow control valves. Such valves can be installed in association with shower roses and taps, where the flow rate can be specified to between say 3 L/min and 20 L/min. (Flow control valves are not restricted to use with showers and conventional taps.) Many types of pressure balancing hot and cold water mixing valves are available for use specifically in conjunction with showers. This is not, however, what we mean by pressure balancing, for this is an 'end point fix' of a systemic problem, which is the subject of our concern. Even so such valves are effective and a limited local solution to the broader matter we are discussing. (The principles of pressure balancing a system are explained in Sturman, 1973). The pipelines installed in many commercial premises will be such that 'end point fixes' are an attractive economical way to proceed. Full pressure balancing of a system will always result in a reduction of the flow available at outlets and appliances and frequently in a saving of water usage.

10.3.2 Water leaks, appliances and water usage reduction

10.3.2.1 Leaks

Leaks in pipes are very hard to detect or to be aware of until exceptional water consumption is found on water bills. High pressure chilled water and especially hot water lines have a potential for developing leaks. Those leaks associated with appliances present somewhat less of a challenge. For example taps can be seen to be dripping, and careful examination of the still water in a pedestal pan will reveal small disturbances to the water surface when water is leaking from the cistern. Alternatively dye can be placed in the cistern and in due course it will become visible if there is a leak from the cistern. By contrast the detection of leaks from underground pipes is a specialised art. We recommend reference to American Water-Works Association (1999) for a thorough treatment of the topic.

10.3.2.2 Refrigerated air conditioning systems

There are air-conditioning standards for interior comfort of refrigerated air-conditioned spaces (see ASHRAE 1997). From a water auditor's point of view we recommend that thermostats be set at the lower extreme of the comfort zone in winter and the higher extreme of the comfort zone in summer so as to save both energy and water. For example setting the interior temperature higher in summer reduces the feed-water required for the cooling tower.

10.3.2.3 Evaporative air conditioners

In a typical evaporative air conditioner the blowdown rate to maintain water quality and thus the water supply feed-rate is approximately 6 to 8 L / hr. The water auditor will often be rewarded, by checking these rates. When the blowdown rate is correctly set to reduce salinity and scaling without excessively using water the blown-down water can usually be used for garden needs, so reusing otherwise waste water. Some modern evaporative air-conditioners have their own controller, which dumps water periodically, and backflushes the evaporation panels.

10.3.2.4 Hot water systems (as distinct from boilers and calorifiers)

We can categorise hot water systems generally as gas, electric, solid fuel systems (storage systems or instantaneous systems), solar systems or heat pump systems. Generally gas systems are to be preferred over electric systems as high-grade electric energy is not wisely degraded to relatively low-grade thermal energy. Often gas is cheaper too, while providing a bonus for reducing greenhouse gas emissions if the electricity is sourced from a coal fired power station. Solid fuel systems (such as wood fired) can be greenhouse neutral if replacement wood is grown locally to match the wood burned. Solar systems are inherently greenhouse friendly, though with a substantial payback period. Heat pump systems are energy efficient, though they tend to be very expensive at the smaller end of the range, with a lengthy payback period. Of all these types of hot water systems there are some common factors to consider.

Among these factors, the location of the unit is important, to minimise heat energy losses through pipe-work, which is best lagged. A temperature setting of approximately 60 deg C for hot water systems is a reasonable compromise between inhibiting fungal and bacterial growth, while not overheating too much the water which is frequently used at about 40 deg C. In many countries the energy efficiency of a unit is identified. While the water auditor audits premises as found, in the suggestions for implementing a water use efficiency strategy, s/he might have the opportunity to specify energy efficient replacement systems, which will thus have a reduced payback period. *Instantaneous heaters* are intrinsically more efficient than storage devices. Importantly they *often have a flow regulator associated with them, and thus specification of other flow control devices (viz flow regulators and some showerheads) downstream can be dangerous.*

10.3.2.5 Showers and taps

High water use efficiency showerheads are available, which produce smaller droplets and have a higher water use efficiency than conventional showerheads. The disrepute into which some early models fell is a barrier to overcome in promoting their use. From a water auditor's point of view, these heads are highly desirable and typically have a short payback period, with the potential of saving around 25 L/min. of water per shower (see above, Table 10.1). *Such showerheads might not be appropriate to install in conjunction with flow-controlled instantaneous hot water systems.* The specifications of the unit and/or the manufacturer should be consulted.

Likewise the installation of flow control devices on hot water taps might not be appropriate in conjunction with instantaneous hot water systems. In many situations, eg. hand basin taps, a cheaper and equally effective way of achieving the same end is to install aerators on the tap. There are also higher cost solutions to saving water at the tap. Quarter tun taps enable almost instantaneous attainment of the flow rate required, without the waste of water while moving the tap through several turns. People with limited mobility of the hands can find them easier to operate. At a higher level of sophistication, single handle or single lever mixing taps are both helpful to people with limited mobility and save the water which would be wasted while separate taps are being adjusted for the desired water temperature. We add the obvious point that no water savings will accrue when a tap is merely used to fill a set volume.

10.3.2.6 Washing machines

Domestic washing machines can be large users of water, using between 90 and 210 L/load, with even greater water use rates for commercial machines. Front loading, horizontal axis, washing machines are much more efficient in their use of water than conventional top-loaders. Recent top loading, horizontal axis, machines are more acceptable to many users. Machines that have suds savers save further water use as well as washing powder.

10.3.2.7 Dish washers.

Dishwashers generally use more water than hand washing. It follows that where a water auditor has an opportunity to recommend dishwashers, only the most water efficient models should be recommended.

10.3.2.8 Dual sinks

Where dishes are washed by hand, the use of dual sinks saves water, for rinsing dishes under running water can be avoided easily.

10.3.2.9 Urinals

Urinals with infra red censors and an associated microprocessor can result in large water savings. Such urinals can be programmed to operate with a varying schedule of use and flushes, but we give a typical example of such a sequence. After a period of non-use, the sensor will detect people's presence and remain latent. Thus non-active visits are discounted. After 5 seconds of an active presence, a wetting pre-flush will occur, and following a single use with a subsequent delay, or a sequence of uses (say four) the processor will flush the urinal fully and then restart the sequence. The cost effectiveness of these units is attractive. Waterless urinals, which use a chemical cartridge are also available.

10.3.2.10 Cisterns associated with water closets

Dual flush cisterns are now widely used. They usually enable the user to select a flush after urinating of half that for defaecating. The effect is to give water savings of around 25% over single flush units and still more over old cisterns. Conversion kits are available in some countries to convert single flush cisterns to dual flush. Such an approach is usually much more cost effective than installing purpose built dual flush devices, though the latter saves more water.

10.3.2.11 Standards

In some countries the national standards association sets standards for various levels of water efficient taps, shower-heads, washing machines, dishwashers, and cisterns.

10.4 ENERGY SAVINGS

When hot water usage is reduced by means of a water efficient shower-head (for example), the energy formerly required to heat the saved water is saved energy. This energy saving is of great interest to the water auditor. It results in financial savings that increase the attractiveness of implementing the water saving recommendations of the auditor, as well as saving energy as a worthwhile pursuit in itself. The energy saved is calculated by simply extending the meaning of the specific heat of water (C_p J/kgK). The specific heat is the amount of energy in Joules (J) required to heat one kilogram (kg) of the water through one degree Kelvin (K). It is relatively independent of the absolute temperature of the water for our purposes. It follows that the quantity of energy (or heat) (Q J) required to heat V Litres (L) of water with a density (ρ kgm^{-3}) from T_l degrees C to T_h degrees C is given by

$$Q = V\rho C_p (T_h - T_l). \tag{10.1}$$

This relationship is not exact, as it assumes that all the energy supplied in the heating process is actually transferred to the water and remains in the water, rather than escaping to the surrounds. Thus an efficiency factor (η) must be introduced into (10.1) to allow for heat losses. This results in

$$Q = \eta\rho V C_p (T_h - T_l) \tag{10.2}$$

The units of Q, in equation (10.2), Joules, will sometimes be inconvenient. Often MJ (10^6 Joules) or kWh (10^3 Watt Hours) will be needed. To convert a value in MJ to kWh, multiply the value in MJ by 0.278 to convert it to kWh. A Watt is a Joule/second. We assume that water has a density of 1 kg/L, which is sufficiently accurate for most water auditing purposes. A typical value of T_h might be 60 deg C, a compromise between having the hot water at a low enough temperature to mitigate scalding while being at a high enough temperature to mitigate algal or bacterial growth. T_l is properly given the mean annual value of the cold water entering the hot water system. In the case study below we assign it the value 16 deg C. The appropriate value will depend upon the location where the audit is being conducted.

Energy savings are also possible by ensuring that the hot water pipes are insulated and that the quality of the insulation is of a high standard. Careful placement of hot water systems or boilers near the end points of hot water use also saves energy. The energy efficiency of hot water systems or boilers should be as high as possible.

10.5 CASE STUDY

A household of four people, two adults and two children, have requested a water audit of their house. They live in a house with one bathroom, one toilet and one laundry. A scheme water service-provider supplies all water used through one metered inlet to the house. The case study will be confined to interior water use in the house only. The electric hot water system is aging and the householders are considering changing to a gas hot water system. How much, they asked, are the operating cost savings for water, gas or electricity if they implement a water-saving retrofit, while either retaining their electric hot water system or changing to gas? What are the payback-times for the total water saving retrofit?

We determined that water costs 60c/kL and electricity costs the householders 13c/(kWh), while natural gas costs 2.1c/MJ. Metered water use for the whole site was obtained from the water service provider's bills over a one-year period - 358 000 L/yr. Then we constructed a water flow diagram, presented in Figure 10.1. This is a little pedantic for such a small water audit, yet it illustrates the procedure. We obtained the flow rates for each water stream. The shower used 16 L/min. and each person showered for 4.0 minutes on average. The bathroom basin tap used 9.0 L/min. at the usual setting, while the laundry trough and kitchen sink taps each used 16.0 L/min at the normal rates of use. All three taps were used for 2.0 minutes per day, we discovered. The current cistern in the toilet uses 11 L/flush and was flushed by each person four times per day. Notice that an estimate of the out-door water usage was necessary to attempt closure, but in this case study we do not discuss how we obtained its value.

Flow rates from taps were determined using both a calibrated container and a stopwatch. The cistern discharge volume was estimated by turning off the supply tap, marking the water surface in the cistern, flushing it, and then filling to the initial water surface with a graduated water container. Members of the family cooperatively measured shower duration times over two days and we took the average. We compared it with typical data for the area (Table 10.1) and the agreement was remarkably good. Water use for washing clothes was determined by the family's rates of use and by reference to the washing machine manufacturer's specifications. From our observations and measurements we constructed a water flow diagram, entering on it the individual flows. (See Figure 10.1. We will discuss this figure further below.) The sum of water usages of unit operations was measured or estimated as 341900L/yr. The usual tolerance for determining closure is 10%. Thus our criterion for closure is (see Figure 1.1):

$$\frac{\Sigma\,(\text{water input quantities}) - \Sigma\,(\text{water output quantities})}{\Sigma\,(\text{water input quantities})} \leq 0.1$$

that is

$$\frac{358\,000\ \text{L/yr} - 341\,900\ \text{L/yr}}{358\,000\ \text{L/yr}} \leq 0.1$$

or

$$0.045 \leq 0.1$$

which is true and so closure has been obtained. In this instance closure is remarkably accurate.

Having obtained closure, we sought to determine a water management strategy. We discussed the performance of the water outlets and especially the shower with the householders and found that they had no problems with water outputs, or shower temperature, when other outlets were in simultaneous use. Therefore we discounted balancing the system or installing a water-balancing valve at the shower outlet. We noticed also that the washing machine was a water efficient horizontal axis machine, which was perfectly adequate. On the other hand the taps in all cases had no flow restricting devices attached and were old standard taps. The shower head was also an older style, high flow rate type. Here we noticed real potential for water savings at low cost. Also we noticed that the toilet had an old cistern, 11 L/flush; giving more opportunity for savings in water usage. We recommended the installation of the devices shown in Table 10.2.

Note that we have recommended aerators rather than flow restrictors, because aerators are very much cheaper than flow restrictors and perform the same task. The cost of purchase and installation of the recommended devices is $517.

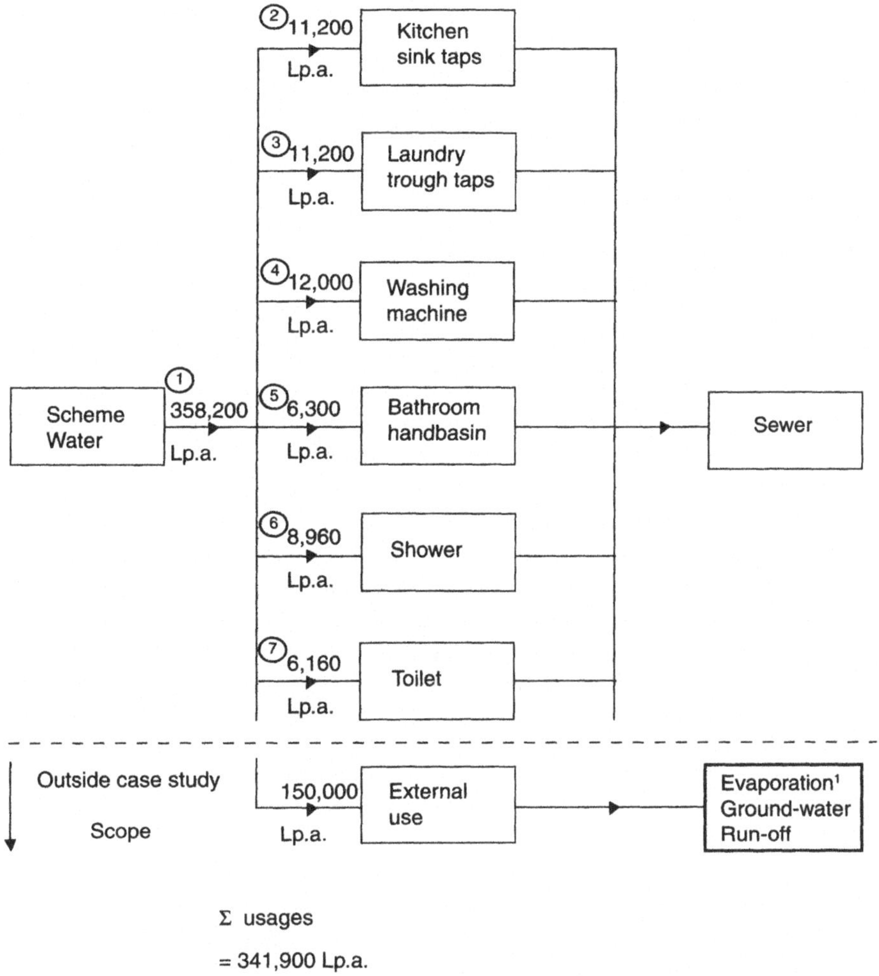

Figure 10.1 Water flow diagram for case study prior to implementation of options for saving water.

There are now several calculations to do. The first is to calculate the water savings due to the installation of the devices recommended in Table 10.2. The second is to calculate the energy savings due to the reduced volume of hot water used. The third is to calculate the savings in annual cost of both water and energy. Finally the capital cost of installing the water saving devices is compared with annual operating cost savings to result in payback time(s) for implementing the recommended water saving strategy(s). We will present here just sufficient of the procedures to illustrate all the necessary calculations.

Table 10.2 Tabulated summary of flows and recommendations forming part of the water management strategy.

Water stream number	Stream description	Measured flow rate	Recommendation	Projected flow rate
1	Mains flow meter	358000 L/yr	Not applicable	255800 L/yr
2	Kitchen sink taps	16 L/min	Flow aerator	10 L/min
3	Laundry trough taps	16 L/min	Flow aerator	12 L/min
4	Washing machine	12000 L/yr	Retain present machine	12000 L/yr
5	Bathroom basin taps	9 L/min	Flow aerators	4 L/min
6	Shower	16 L/min	Water efficient shower head	9 L/min
7	Toilet	11 L/flush	Dual flush cistern	6 or 3 L/flush

The shower's present water usage per annum is given by: 16 L/min x 4 mins/shower x 4 showers/day/person x 4 persons x 350 days/year = 89600 L/yr. The shower's potential water use with the new shower head installed is calculated as above with 9 L/min replacing the former 16 L/min resulting in 50400 L/yr. The anticipated saving in water is therefore 39200 L/yr. The savings for kitchen sink taps, laundry trough taps, bathroom basin taps and toilet are respectively 4200 L, 2800 L, 3500 L and 36400 L. Thus the total water savings per year are 86100 L. The total hot water savings we take to be half the savings of all but the toilet (this proportion might be slightly different in some situations) that is 24900 L.

Using equation (10.2) the total energy savings are: 1350 kWh for the electric hot water system (using η =0.95) and 1630 kWh for the gas hot water system (using η = 0.8).

The saving in annual cost for water is 86.1 kL x 0.60 $/kL = $52.

The saving in annual cost for energy is 1350 kWh x 13 c/kWh = $175/yr (electricity) and

1630 kWh x 2.1 c/MJ / 0.278 kWh/MJ= $123/yr (gas).

The payback period for implementing the water management strategy is $517/($175/yr + $52/yr) =2.28 yrs for the electric hot water system and $517/($123/yr + $52/yr) = 2.95 yrs for the gas hot water system.

Note that we have not included the cost of the new gas hot water system in the calculation because the householders were considering changing hot water systems independently of the water audit.

Finally we make some observations on the case study. Because the water audit is so simple we have followed the procedures outlined in Chapters 3 and 4 in simplified form. The water flow diagram raises some interesting issues. Notice that no flow measurements have been taken on the outlet sides of the unit operations, or of the sewer flow. This is consistent with our earlier comments in Chapter 3. Strictly the water audit of the case study is only a partial audit, though consistent with normal practice for internal commercial audits. Auditing the outflows from the unit processes and the sewer would constitute a waste-water audit as indicated in Chapter 3.

In this simple audit the water flow streams correspond with actual pipes. In a more complex audit this would almost certainly not be the case. The flow diagram in Figure 3.2 is an example where the water flow streams are aggregations of physical pipes in some instances.

10.6 REFERENCES

American Water-Works Association 1999. *Water Audits and Leak Detection M36,* Hanover.

ASHRAE 1997. *ASHRAE Handbook – Fundamentals*, American Society of Heating Refrigeration and Air-conditioning Engineers, U.S.

Sturman, J.J. 1973 A method of designing multi outlet vacuum systems served from central rotary vacuum pumps, *I. E.Aust. Mech. & Chem. Engg. Trans.*, **MC9**(1&2), 20-24.

Wajon, E, 2000. Water usage and water efficient options for staff amenities. Kinhill Pty Ltd, Western Australia. In: Murdoch University (2001) *Principles of water conservation and water auditing unit guide.*

Water and Energy Savers 1994. *Participant's handbook - Water efficiency assessors training course (inside water use).* Master Plumbers and Mechanical Services Association of Western Australia, Perth.

11

Arena: External commercial water use

11.1 INTRODUCTION

By external commercial water use we mean water use outside of commercial premises, that is, on lawns and gardens associated with commercial buildings. Agricultural use falls outside the scope of this chapter. Lawns associated with sporting facilities and municipal parks have much in common with the topic of this chapter, yet we will defer the consideration of some detailed issues relating to turf and irrigation until the next chapter.

In general the water auditor takes what s/he finds and makes suggestions for modifications in the water management strategy. Sometimes the inefficiencies of existing irrigation systems and external landscapes are so glaring, and result in such high operating costs, that a complete re-design is warranted. Other professionals might need to be involved in implementing such a process, but to initiate the process in the water management strategy, the water auditor needs a sound understanding of the principles involved. With this in mind the contents of this chapter appraise the external landscape associated with commercial premises with a view to enabling the water auditor to make astute recommendations in the water management strategy.

We make the observation that all of the external grounds of a commercial building or complex need not be covered with vegetation. Some ground will likely include surfaced roads. Concrete or brick paving might cover significant areas, often transition areas between the building itself and vegetated ground further away from the building. Paved spaces are often used for outdoor entertainment or heavy human traffic. Clearly paving requires no use of water and even raises the possibility of harvesting run-off, which might become a source for partial irrigation of vegetated areas. There is also the possibility of substituting paving for vegetation over some areas to reduce water requirements. Disadvantages of doing so include the capital cost and the possibility of higher local temperatures (or colder local temperatures) due to the paving absorbing more heat from the sun by day (or radiating more heat to the clear sky by night). Over those areas where vegetation is used, the major factors we consider are hydrozones, dimensioning, vegetation types (including lawn types), windbreaks, mulching, soil amendment, irrigation operation, waste-water reuse and water resource substitution.

11.2 PRINCIPLES OF LANDSCAPE DESIGN FROM THE WATER AUDITOR'S VIEWPOINT

11.2.1 Hydrozoning

Hydrozones are zones within each of which the vegetation requires a similar amount of water. The hydrozone is the result of purposeful selection of plants with similar water requirements, resulting in greatly simplified irrigation design. Thayer & Asler (1984) present a matrix that relates the placement of hydrozones to human contact, visual importance and to water and energy needs (see Table 11.1). The result of following the scheme of this table is to limit the use of water, with the highest density of water use located in those areas with maximum human contact and visual amenity. Further away, water usage decreases, reaching zero in the elemental zone.

Table 11.1. Hydrozones related to human contact, visual importance and to water and energy needs. (Adapted from Thayer & Asler, 1984)

Hydrozone	Human contact	Visual importance	Water and energy needs
Primary	Direct, intense, active	Conspicuous	Greatest
Secondary	Less: direct, intense More: passive	Conspicuous	Reduced
Minimal	Little	Less conspicuous	Slight
Elemental	None	Inconspicuous	None

Decorative, high water use, exotic and local plants, together with turf of the highest visual amenity and receiving the heaviest human traffic, are located in the primary hydrozone. The secondary, minimal and elemental hydrozones will have plants that use progressively less water and turf of reducing quality appearance and ability to handle traffic. Within each hydrozone the vegetation is chosen so that plants all remain healthy when the irrigation uniformly applies the same amount of water with the same controller schedule.

11.2.2 Dimensioning

The hydrozone principles applied in isolation from the choice of dimensions of the hydrozones do not guarantee a satisfactory design from the point of view of surface (sprinkler) irrigation. Where surface sprinklers are used, the dimensions of the grassed areas need to be integral multiples of the sprinkler effective half-throw. Small narrow strips of lawn are an irrigation designer's nightmare. Generally they cannot be accommodated efficiently in the irrigation zoning of adjacent shrubs or other plants. Sprinklers at the straight edges of lawns are properly specified with 180 degree heads, those at right angled corners with 90 degree throw heads. The constraints we have outlined do not apply when sub-surface trickle irrigation is used. We discuss types of irrigation below.

11.2.3 Vegetation types

11.2.3.1 Plants

Local plants are almost always to be preferred over exotics, unless there are specific reasons for including exotics. Such reasons might include lower water requirements, seasonal shade provided by deciduous plants or the aesthetics of the plants' flowers. In some circumstances high water use plants might be desirable to control shallow water tables. In climates where there is rain for much of the year, or where plants do not need irrigating, the water auditor's interest in the external grounds to a commercial enterprise will be lessened. Therefore we will concentrate on vegetation types for use in drier and warmer climates.

Plants that require less water utilise a range of adaptations including small leaf areas, leaves arranged in special ways, or even chlorophyll surfaced stems and very reduced true leaves. Some such plants (eg. *Acacia acuminata*) require no watering, and so would be assigned to the elemental hydrozone. Plants that require some watering will be assigned to the appropriate hydrozone that is to minimal, secondary or primary hydrozone. The increase of water use as we move through these hydrozones has been used in some countries to classify plants on the basis of their increasing need for water. For example a plant labelled 'one drop' might correspond to the minimal hydrozone, while one labelled three drops might correspond to the primary hydrozone. While schemes can vary according to different countries, an example of such a scheme in Western Australia can be found at the web site: http://www.watercorporation.com.au/environment/Plants/index.html. The Water Corporation of Western Australia is unique in being responsible for water supply over the largest surface area of land of any water service provider in the world. Though the population supplied is not large, the area of land for which it is responsible has warm climates, ranging from Mediterranean, through dry sub-tropical, seasonally wet tropical to hot dry desert. Thus the corporation has experience over many of the climatic types where water conservation issues will be particularly important for the water auditor. The web site gives a comprehensive list of plants arranged according to one, two or three drop water usage. Other information on garden design is available too; see http://www.watercorporation.com.au/savewater/index.html.

11.2.3.2 Lawns

Lawns often account for a large portion of the external grounds of commercial premises. Thus their water needs will attract considerable attention from the water auditor. Grasses are divided into two broad categories: warm season grasses and cool season grasses. Examples of the former are Bermuda Couch, Buffalo, Kikuyu and Saltene. Examples of the latter are Bent grass, Ryegrass, Tall Fescue and Kentucky Blue. Colwill (2001) presents a table, which relates grass type with water use, drought tolerance and other parameters. We present the table, with modifications, in Table 11.2.

Notice that cool season grasses generally use more water than warm season grasses. On the whole their other characteristics compare less than favourably with warm season grasses. Thus in warm climates, where irrigation is needed for lawns, we recommend the use of warm season grasses.

Table 11.2. Water use by warm and cool season grass varieties, drought tolerance and other parameters. Adapted from Colwill (2001).

Type of grass	Water use	Drought tolerance	Fertiliser requirement	Heat tolerance	Shade tolerance	Wear resistance
Warm Season						
Bermuda couch	Low	V. high	Med/high	Excellent	Poor	V. poor
Santa Ana	Low	High	Med/high	Excellent	Medium	Good
Greenlees Park	Low	High	Med/high	Excellent	Fair	Fair
Wintergreen	Low	High	Med/high	Excellent	Medium	Good
Windsor Green	Low	High	Med/high	Excellent	Medium	Good
CT-2	Low	High	Medium	Excellent	Fair	Good
Saltene	Medium	Medium	High	Excellent	Medium	Good
Buffalo	Medium	Fair	Medium	Excellent	Good	Fair
Kikuyu	High	Good	Low	Excellent	Poor	V. good
Cool Season						
Kentucky Blue	V. high	Medium	Med/high	Fair	Poor/fair	Fair
Bentgrass	V. high	V. low	High	Fair	Good	Poor
Ryegrass	V. high	Low	Medium	Poor/fair	Fair	Good
Tall Fescue	V. high	Med/high	Medium	Good	Fair	Good

11.3 STRATEGIES FOR REDUCING WATER USE

11.3.1 Irrigation types

The water auditor starts with the irrigation type s/he finds. Generally this will reflect past practice and therefore will often include above ground sprinkler irrigation. Such irrigation has many disadvantages, which include evaporation loss through misting, wind drift, increased likelihood of fungal diseases in plants and slippery paths to mention just a few. Within the development of a water management strategy the water auditor has two clear choices: to optimise an existing sprinkler system or to recommend the installation of below surface drip irrigation (lawns or plant beds) or sub-mulch irrigation (plant beds). The auditor will evaluate these options in the water management strategy. Advantages in below ground irrigation, or below mulch irrigation, include being relatively vandal proof and saving water usage by at least 25% and possibly by up to 60%, resulting from almost no evaporative losses, no over spray and no windage losses. While sprinkler irrigation ideally takes place between approximately 4am and 8am, subsurface trickle irrigation can be applied more flexibly. This can result in reduced sizes of feeder mains and manifolds, with associated reductions in costs compared with a sprinkler system. (Unless the water auditor has specific training in irrigation design, s/he needs a specialist to assist with detailed design and evaluation.) On some sites future development might include further garden or lawn landscaping. In such a situation the costs of a sprinkler system and a below ground drip system can be compared directly and the best choice made prior to installation.

11.3.2 Windbreaks

The principle of the windbreak is that transpiration and convective losses of water from wet plant leaves and ground surface (sprinkler irrigation) or from transpiration (below mulch or below ground irrigation) are reduced as the wind speed on the surface of the plant leaves and the ground is reduced. Windbreaks are located to the windward side of the area where protection is sought, ideally with the major axis of the windbreak vegetation at right angles to the direction of the mean prevailing warm season wind. Sometimes a windbreak can be valuable in protecting vegetation from cold, dry cool-season winds. A good qualitative discussion on windbreaks can be found in Mollison & Slay (1994). The plants constituting the windbreak should be selected for four factors: tolerance to wind, speed of growth, attainable height and permeability. We further discuss the latter two.

If the horizontal extent of the vegetated ground to be protected is known, the height of a permeable windbreak is determined by a rule of thumb to be approximately one twentieth of the horizontal extent. Even so, the windbreak will afford some minimal discernable reduction of wind speed out to as far as thirty windbreak heights downwind of the windbreak. The structure of the windbreak consists typically of three parallel rows of shrubs or trees. The central row would be staggered with respect to the two outer rows. Where the windbreak plant is itself permeable because of spaced branches or sparse foliage, the plants can be planted relatively close together: that is, the drip lines might almost touch. On the other hand, if the trees or shrubs have dense packing of branches and leaves, typically half to one plant width is left between the drip lines of the mature plants. At the start of growth of the plants constituting the windbreak, they will be short, and the spacing between plants can be appropriate to their immature dimensions, with the number of rows increased. As the plants grow, alternative plants can be cut to increase the permeability again as the individual plants spread. Our reason for promoting permeable windbreaks is that their efficacy extends far further down wind of the break than is the case for relatively impermeable windbreaks, where the break will be effective for only a few plant heights down wind.

11.3.3 Mulching

Mulch is a coarse substance of a variety of shapes and sizes, which is placed on the surface of the soil in the vicinity of plants. The mulch can be inorganic, like gravel, or organic like wood chips. It is essential that the mulch does not absorb water to any extent, and that it has a permeable nature, which admits water so that it drains freely. Organic mulch is preferable to inorganic mulch because it breaks down over time, increasing the organic content of the soil and enhancing plant nutrient uptake. We caution that the process of breaking down removes nitrogen from the soil, and it needs to be replaced periodically. If the organic material readily absorbs water it ceases to be effective as mulch, as it becomes a transport mechanism for water from the ground to the atmosphere above the mulch. Inorganic mulch, such as gravel, is to be preferred over organic mulches that absorb water readily. The mulch saves water by inhibiting water transport from the soil surface to the atmosphere. Apart from this there are other benefits from mulching.

Among these benefits we note that plant roots are protected from ambient temperature fluctuations, erosion by either wind or water is inhibited, normal soil biological activity (bacteria, earthworms) is enhanced and weed growth is greatly reduced. Furthermore the placement of mulch allows trickle irrigation tubing, that would otherwise need to be buried, to be laid out on the soil surface and then covered with mulch in many instances, saving greatly on installation costs. The benefits of the mulch are not fully realised unless it is sufficiently thickly applied. Opinions vary on the minimal desirable thickness. In our opinion it is 75mm or even more. Mulch does not benefit the stem of many plants and can lead to collar rot. Mulch is therefore best kept at least 50 mm away from stems. Other maintenance matters need attention too.

If an organic mulch is used it will slowly disintegrate. Therefore in order to maintain its benefits we need to top it up to at least 75mm thickness on a regular basis. In the process of topping up the existing mulch must not be grossly disturbed. If it is, the minute plant roots that develop at the soil-mulch interface will be damaged and in turn so will be the plants.

11.3.4 Soil amendment

Loam soil provides the maximum moisture availability at approximately 17 mm per 100 mm depth of loam. By loam we mean a soil constituted by approximately 40% sand, 40% silt and 20% clay. (Clay has particles less than 2 microns, silt particles between 2 and 50 microns and sand has particles more than 50 microns in diameter). The mix of particle sizes represented by loam has an intermediate value of hydraulic conductivity and will receive an intermediate amount of rain prior to run-off starting. These are desirable features. A soil with a high hydraulic conductivity or permeability will quickly transport water below the root zone of plants, while one with low hydraulic conductivity will induce run-off of rain (for sprinkler irrigated water) while the water has yet to penetrate to the limit of plant root depth. Loam balances these requirements, while providing maximum moisture storing capacity. The purpose of soil amendment is to move an existing soil mix in the direction of the optimum loam by mixing with it soil with characteristics which compensate for the existing soil's deficiencies in balance of particle sizes. Here we will not pursue soil amendment techniques in depth. We simply indicate that a sandy soil might be amended by the addition of a silty clay, and a heavy clay soil might be amended by the addition of a silty sand. This is physical soil amendment, yet soil amendment means more.

Normally top soils are characterised by the presence of relatively high amounts of organic matter. This organic matter further enhances water-holding capacity, also aids the uptake of nutrients by the micro-roots of plants and provides a friendly environment for earth-worms who bring associated benefits. Thus soil amendment can imply not merely the physical amendment described above, but also organic amendment. The addition of animal manures or organic fertilisers and mixing them with the soil constitutes organic amendment, which is subsequently aided by organic mulch above the soil. Soil amendment takes place prior to planting not after planting, for after planting, the mixing processes destroy plant micro-roots. The soil amendment is applied to the top zone of the soil where plant feeder roots are most active. Typically the top 200mm of soil would be amended.

From the water auditor's point of view soil amendment offers direct savings in water use by reducing either run-off, or infiltration beyond the plants' root zones. Indirectly, we can save water by enhancing plants nutrient uptake, which issues in plants with a healthy appearance. When plants appear healthy, excessive watering is less likely by ground staff, for often over-watering is the first response to plants that look unhealthy for other reasons than lack of application of water.

One reason plants can be unhealthy is that the water being applied does not wet the soil. Some organic materials in soils can coat soil particles with a water repellent waxy substance. Soil wetting agents can be applied to enable water easily to wet and to penetrate the soil again. Domestic detergents, or industrial detergents, are not suitable. Irrigation specialists and many nurseries stock purpose designed soil wetting agents.

11.3.5 Wastewater reuse

This topic will be treated in the next chapter. Here we simply flag that we consider that wastewater reuse in the context of external commercial water use is appropriate, especially if it is to be applied below ground. Local government approval might be necessary.

11.3.6 Irrigation operation

Irrigation is a complex process. Irrigation design is a specialist activity, which goes beyond the expertise of most water auditors who are not specifically trained for it. This book is not intended to include comprehensive material for irrigation designer training. Yet some awareness of irrigation design, and particularly operation and maintenance issues, is necessary background for the water auditor to make credible recommendations regarding water savings and irrigation. The level of awareness required is dependent upon the arena of the irrigation. External commercial water use often involves minimal gardens and lawns in the vicinity of buildings in which greater quantities of water are used. In such a case the importance of the water use in the gardens/lawns is relatively small. On the other hand sometimes the major water use of the head office of a major corporation could be the gardens/lawns, which might be extensive and be kept in excellent condition to foster prestige or image. As a generalisation we believe that external commercial water use is to be distinguished from sporting and municipal complexes in that the major purpose of the site is usually not the gardens. Therefore we will provide comments about irrigation that are of the nature of rules of thumb for good practice. We defer more rigorous treatment until the next chapter.

Generally speaking an automated irrigation system can water a garden more efficiently than a hand or sprinkler based manually operated system. Usually, therefore, the water auditor is on safe ground to recommend the installation of an automated irrigation system to replace a manual system. Sprinkler systems are still extensively used, though below ground drip systems have many advantages.

When a sprinkler system is installed, the sprinkler spacing is best at one radius of the sprinkler throw. Thus water from one sprinkler just reaches the adjacent sprinkler(s). The uniformity of watering will usually be satisfactory with this arrangement. Droplet size is an important parameter. A large droplet size will lose less water to evaporation and less water will be displaced from its intended range by wind (wind drift). Thus sprinklers are best chosen for large droplet sized sprays. As time passes the nozzles on sprinklers deteriorate and seals can deteriorate also. Replacing these can improve the performance of the sprinkler irrigation system and lessen water usage.

The operating start time of the sprinkler irrigation, the period of time over which the water is applied and the frequency of applications are of great importance. We give examples here of rules of thumb which we have found adequate, though sub-optimal. An industry rule of thumb is that the same quantity of water is applied to each hydrozone, but the frequency of application is varied to give the correct amount of water. How much water do we apply? Take the highest water use hydrozone. Apply 10 mm of water per application, and apply it at a rate that does not exceed the following values: 50 mm/hr for sandy soils, 13 mm/hr for silt loam and 3.8 mm/hr for clay. These watering rates are for bare soils on slopes of less than 5%. With cover, the latter two rates can be increased to 25 mm/hr and 5 mm/hr respectively. The frequency of application of the water is such that averaged over an arbitrary but lengthy period (say two weeks) the average daily application rate is 60% of net pan evaporation rate, assuming no rain. Hydrozones requiring less watering have the same application at a reduced frequency, proportional to the vegetation's needs. Returning now to the irrigation start time, note that sprinklers never should operate in the middle of the day, as up to 50% of the water applied can be lost to evaporation in warm dry climates. Application of the water late in the afternoon, or in the evening, can be accompanied by fungal problems, especially if evenings are warm. We recommend that sprinklers apply the water in the morning, between say 4am and 8am. Water will infiltrate the soil so that the plant(s) then has access to it when it is needed during the day.

With below ground drip irrigation, the above requirements can be relaxed. Water usage can be assumed to be at least 25% less and the time of starting application of the water is freed from constraint.

There are methods for reducing water use by overriding the irrigation controller when it rains. One method is to install a water-activated switch in a roof gutter, interrupting the common line from the irrigation controller when a predetermined water level is reached in the gutter. Another method is to use a soil moisture sensor to shut down the irrigation controller when soil moisture reaches a predetermined level. Manual override is a third, but less desirable method, as human intervention, which starts with good intentions, can lapse over periods of time.

11.3.6.1 Minor case study

Sprinkler irrigation is to be applied to an area in the primary hydrozone, with good plant and mulch cover. The net pan evaporation is 9 mm/day, the soil is sand and the surface is flat. What is the frequency of application of the irrigation at 10mm per application? What is the shortest period of time over which the 10mm of water should be applied? How much water would be saved if the irrigation were applied by sub-soil drippers and the area watered were 200 square metres?

Over a period of two weeks a net pan evaporation rate of 9mm/day invites the total application of 0.6 x 9 mm/day x 14 days = 76 mm of water. 10 mm / application of water therefore implies that the number of applications required over the two weeks is 76 mm of water / (10 mm / application) = 8 applications (rounded to the closest integer). The period of application is 14 days / 8 applications = 2 days / application. (Rounded to the nearest integer). We now check our resulting irrigation schedule (10 mm per 2 days) against our initial requirement. There will be 8 applications x 10 mm / application = 80 mm provided per fortnight. This is slightly more than required, but it is acceptable.

The maximum application rate for sand, whether covered or uncovered, is 50 mm/hr to ensure no surface run-off. To reach or exceed this application rate, 10mm of water would need to applied in 10 mm / 50 mm/hr = 0.2 hours or 12 minutes. Run-off will not be a problem if the sprinklers deliver 10 mm of water over a longer period than 12 minutes.

To calculate the water saved if below ground drip irrigation were used we take account of the reduced requirement for application of water of at least 25%. Thus the potential water saving over 200 square metres is 0.25 x 10 mm x 200 m^2 / 2 days = 0.25 m^3/day or 250 Litres per day or more.

11.4 WATER RESOURCE SUBSTITUTION

The water auditor is concerned not merely with reducing the volume of water use, but also with attempting to source water of appropriate quality for its application. In many commercial enterprises treated scheme water of high quality is used for irrigation purposes. This practice is both expensive and inappropriate if other sources of water are available. Bore water is one possible resource that might replace scheme water. Storm water run-off provides another possibility, which can be especially valuable in warm or hot dry climates. Run-off from the site can be collected in shallow basins, cascading to further basins, with each basin surrounded by vegetation. The basins themselves can be small, or large enough to be turfed areas used for recreational purposes for the majority of the time when they are empty of water. The main aim of the cascading basins is to increase the soil moisture content and retain it for longer periods in pre-selected locations where vegetation is required. Hill & Nicholson (1989) provide a brief description of a design process for such basins. Rainwater collected from roofs is another possible resource for water. On the whole it is expensive water because of the cost of the storage tank(s), and its quality is greater than necessary for most irrigation purposes. In some hot dry climates and remote areas this might be a suitable option.

11.5 REFERENCES

Colwill, J. (2001) Plant, tree and turf selection. In: *N230 Water conservation and water auditing.* Unit Guide 2001, Division of Science and Engineering, Murdoch University, Western Australia.

Hill, A.L. & Nicholson, C.J. (1989) *Water conserving design for gardens and open space.* Water Authority of Western Australia, Perth, Western Australia.

Mollison, W. and Slay, R.M. (1994) *Introduction to permaculture.* Tagari Publications, Tyalgum, Australia.

Thayer, R.L. & Asler, T.R. (1984) Water Conserving Landscape Design. In McPherson, E.G. (Ed.) *Energy conserving site design,* Watkins & Son, Providence, Utah.

Water Corporation of Western Australia.

Plant selection: http://www.watercorporation.com.au/environment/Plants/index.html

Garden design: http://www.watercorporation.com.au/savewater/index.html.

12

Resources: Domestic and small commercial wastewater reuse

12.1 INTRODUCTION

This subject is particularly difficult to deal with in general, global terms, without reference to a particular geographical and politico-regulatory setting. We have therefore referred extensively to Australian settings, which we might regard as a type of case study. Clearly in other settings, local statutory and regulatory particulars will need to be accessed by the water auditor who wishes to investigate the reuse of wastewater in that setting. Domestic wastewater reuse is currently not permitted anywhere in Australia but is widely supported by the community, promoted by researchers, and greywater reuse is improvised by up to 20% of householders. Its widespread implementation will make an enormous contribution to the sustainability of water resources. Integrated with other strategies in the outdoor living environment of settlements in arid lands, great benefit will be derived. This chapter describes six options for wastewater reuse and case studies are given where productive use is being made for revegetation and food production strategies at household and community scales. Pollution

 Water Auditing and Water Conservation by Jeffrey Sturman, Goen Ho and Kuruvilla Mathew. ISBN: 1 900222 52 3. Published by IWA Publishing, London, UK.

control techniques, public health precautions and maintenance requirements are described. New Australian design standards and draft guidelines for domestic greywater reuse introduced by the Western Australian State government agencies for mainstream communities are evaluated. It is recommended that dry composting toilets be coupled with domestic greywater reuse. Various types of systems available in Australia are described. For situations where only the flushing toilet will suffice the unique "wet composting" system can be used and this also is described. A vision for household and community-scale on-site application is presented.

The conventional paradigm governing wastewater management has focussed on the removal of pollutants in the wastewater and disposal as the solution. It relies on centralised water supply, sewerage and drainage systems with up to 85% of costs incurred in piping and pumping. This paradigm was developed on the Thames River in the last century and its appropriateness for the vast dry continent like Australia has been questioned (Newman & Mouritz, 1996) as has the transfer of these expensive centralised systems to developing countries (Niemczynowicz, 1993). Indeed, the arguments for abandonment of this paradigm in favour of one which cycles nutrients and resources for sustainability are perhaps now as evenly matched against the *status quo* as they were in the last century when the 'water carriage' lobby narrowly defeated the 'dry conservancy' lobby (Beder, 1993). The latter then also sought separation at source with reuse of dry and liquid products for agriculture although with much less scientific basis than what is available today. Goodland and Rockefeller (1996) proposed three general principles to enable the passage of the new sustainable paradigm: a) cease expansion of sewers and commence decommissioning them; b) promote on-site recycling systems that avoid pollution of water resources; and c) charge the true value of water. The focus of this chapter is on-site recycling systems.

Reuse of wastewater occurs most effectively with on-site (localised) or small-scale treatment systems. It was not possible to reuse all the effluent from centralised treatment plants in the sewered suburban sprawl as there simply was not enough land for nearby broadacre application. Thus to achieve the goal of total reuse the involvement of a local community in the urban situation would have to be enabled and reuse options in the local context agreed upon. In sewered areas greywater reuse can still be implemented on-site. Greywater or sullage is effluent from the bathroom, washbasin and laundry, and for primary systems should exclude kitchen sink wastewater as it carries oils and high BOD. The more concentrated blackwater (from the toilet) can still go to the sewer along with kitchen effluent. In unsewered areas the blackwater can be treated separately or dry vault (pit or composting) systems utilised. Greywater reuse can result in cost savings (to both the consumer and water service provider), reduced sewage flows in sewered areas and potable water savings of more than 40% when combined with sensible garden design (see Figure 12.1).

Significant impact on water and energy use might require greywater reuse to be coincidental with water-sensitive urban design, reduced lawn area, and possibly the growing of food at home and in public open space. There is immense community support for reuse of wastewaters (WAWA, 1994). This chapter will review regulatory developments, describe six methods of options and explain the broader design approach that needs to be applied with greywater reuse.

12.2 CURRENT REGULATION

Domestic greywater reuse, governed in W. Australia by state and local government health acts, is currently not allowed in any of the Australian States, although WA State authorities acknowledged that 20% of householders engaged in this practice in Perth (Lugg, 1994; Stone, 1996). In Queensland three options were developed for possible implementation (Department of Primary Industries, 1996). The model guidelines for domestic greywater reuse in Australia (Jeppeson, 1996) covered hand basin, toilets, primary greywater systems (direct subsurface application) and secondary greywater systems (mesh, membrane or sand filtration prior to irrigation). For primary systems the guidelines have adopted the Californian approach requiring the use of a surge tank with a screen to remove lint and hair. Electrical power is therefore required for the automatic pump system and weekly inspection and clearing of the screen. The need for maintenance to these components by the householder resulted in some 80% of Californian systems being in an unsatisfactory condition. The recommendation of this approach as the solution is questionable. The updated standard AS1547-1994 (Standards Aust./NZ, 1996) guiding domestic effluent management is significantly more progressive in providing design criteria for a range of treatment systems with reuse, opening the way for further innovation.

Treated effluent from centralised plants is used on municipal ovals, parks and golf courses in many country towns of Western Australia (Mathew & Ho, 1993). In New South Wales (NSW) treated effluent from centralised plants is allowed in urban areas (NSW Recycled Water Coordination Committee, 1993). National guidelines for the use of reclaimed water via dual reticulation have been prepared (National Health & Medical Research Council, 1996). The level of treatment recommended is secondary plus filtration and pathogen reduction. Alternatives to this include constructed wetlands, which may achieve treatment equivalent to open water areas, which will allow pathogenic die-off due to UV sterilisation.

In 1996 the Western Australia Government released its Draft Guidelines for Domestic Greywater Reuse (HDWA, 1996) which allow the public to install greywater reuse systems in three shires as a means of conducting trials for 12 months (Fimmel, 1997). The three shires provided different soil types which would no doubt call for different design responses to pollution control and absorption. Moreover many dwellings in these areas were unsewered.

The Western Australian State Government agencies quite rightly want to respond to the massive public interest in greywater reuse while at the same time exercising caution after the early Californian experience. The three shire trial will provide broad experience if a range of systems are allowed. The monitoring of these will provide invaluable information: Which systems are most appropriate for each of the conditions? How effective are the local government authorities in providing support and direction? How diligent are householders in maintaining these systems? What are the economic benefits? How effectively are greywater systems integrated into the landscape in relation to productivity and nearby recreation? What are the longer term effects on soil and plants? What is the nutrient balance between inputs, plant uptake, and percolation into the soil?

12.3 SYSTEM CHARACTERISTICS

A greywater reuse system needs to protect public health, protect the environment, meet community aspirations and be cost-effective. Current on-site treatment systems have generally adopted the technology of the conventional activated sludge plant for large treatment systems. If removal of nutrients is required for installation of on-site units in nutrient-sensitive catchments, phosphorus (P) can be removed by alum dosing and nitrogen (N) by nitrification and denitrification in separate chambers or by intermittent aeration of a modified activated sludge set-up.

If the effluent is used for irrigation of garden plants there is the question as to why N and P should be removed. There may be an imbalance between plant requirement for the nutrients and the seasons, with a higher requirement in the warmer months than the colder months. Rather than removing the nutrients, an alternative is to store the nutrients in the soil. Soils containing clay have the capacity to absorb ammonium and phosphate present in secondary effluent. Sandy soils can be amended with clay, loam or if convenient the 'red mud', bauxite-refining residue. The most progressive application of domestic greywater reuse appears to be in California. But even there the minimum prescribed depth of 430 mm for subsurface irrigation "ignore(s) the importance of aerobic bacteria and biota (found in profusion in the top few inches of garden soil) for digesting organic matter, nutrients and possible pathogens found in graywater" (Kourik, 1995).

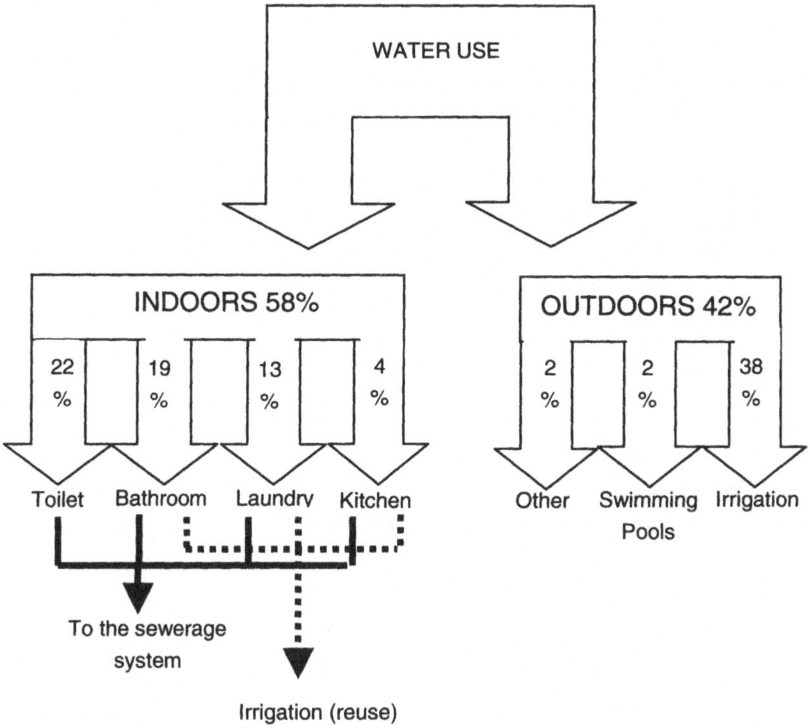

Figure 12.1 General distribution of water use patterns and greywater reuse possibility.

12.4 SIX OPTIONS FOR SMALL SCALE WASTEWATER REUSE

12.4.1 Background

All the six options of localised wastewater reuse involve septic tank systems as the first treatment. So proper design and construction is very important. Septic tanks treat the settled waste anaerobically. They have a retention period of three days. A thick crust of scum formed at the surface helps to maintain the anaerobic conditions. Sludge accumulated has to be removed once in 1-5 years. Effluent has high BOD. The reduction in BOD is in the order of 250-300 mg/L in the influent to 150-200 mg/L in the effluent. Effluent will be high in nutrients with high level of pathogens.

12.4.1.1 Septic Tanks

The following points summarize the function and operation of septic tanks:

- Septic tanks treat the settled solids anaerobically.
- Have a retention time of three days.
- A thick crust of scum formed at the surface helps to maintain anaerobic conditions.
- Sludge accumulated has to be removed once every 1-5 years.
- Effluent has high BOD.
- Effluent will be rich in nutrients and have a high level of pathogens.
- Both sludge and effluent are to be treated and disposed of without human contact.

12.4.1.2 Design of Septic Tanks

For a family of five people using water at the typical rates of a Perth family:

• Water consumption	200L/day
• Total quantity: 200 X 5	1000L/day
• 60% of the quantity ends up in septic tank	600L/day

Septic tanks provide storage for 3 days.

• Volume of tank = 600 X 3	1800L
• Depth of tank not less than 1m	1-1.5m
• Area of the tank: 1.8m^2	
A tank of 1.5m diameter	1.76m^2

Present practice in Western Australia

- Two tanks 1.5 and 1.2m in diameter
- 1.2m diameter has an area of 1.1m^2
- Total area = 2.86m^2
- This will provide for overloading and shock loading.

12.4.1.3 Desludging Frequency

Desludging is necessary when 1/3 tank is full of sludge.

Sludge volume = 1.8/3 = 0.6m^3

Sludge produced per person/year 0.03 to 0.04m^3

Number of years: 0.06/(0.03 to 0.04)*5

= 3-4 years.

If you consider volume of both tanks:

Number of years = 2.8/(3*(0.03 to 0.04))*5

= 4.5-6 years.

Frequency can be reduced by adopting a bigger tank. With two tanks of 1.5m diameter:

Volume: 3.6m^3

Frequency = 3.6/(3*(0.03 to 0.04))*5

=6-8 years.

If nutrients and pathogens are to be reduced, soil filtration can be used.

12.4.2 Amended Soil Filter

The system illustrated in Figure 12.2 is an Ecomax system for domestic wastwater treatment (Bowman 1996). Ecomax consists of a conventional septic tank and dual leach drain or soakwell modified by the addition of a filter bed of amended soil with a plastic lining. The filter bed contain neutralised bauxite residue which has the capacity to absorb phosphorus, ammonia and heavy metals. Phosphorus and heavy metals are removed by adsorption. Because of the fluctuating flow, flooding and drying patterns occur. This produces alternate anaerobic and aerobic conditions creating a possibility of nitrification and denitrification. Nitrogen is removed by this operation.

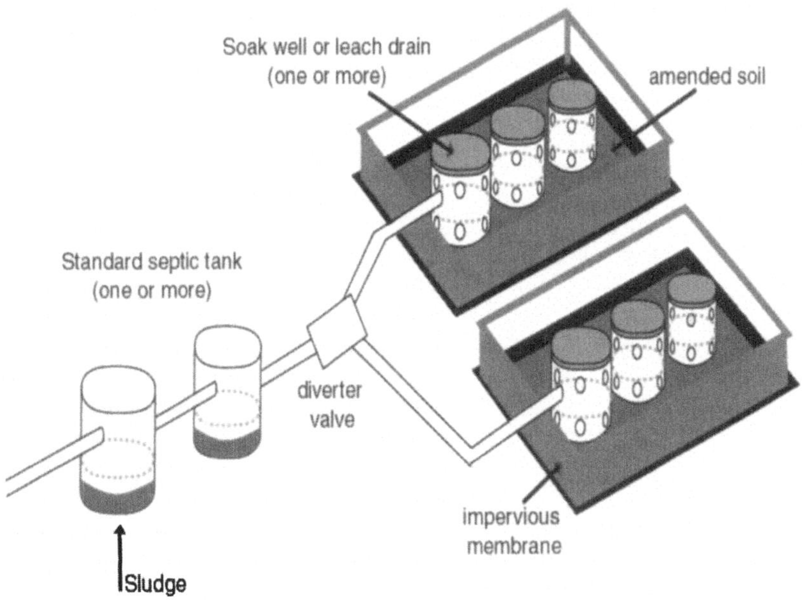

Figure 12.2 Schematic diagram of Ecomax wastewater treatment system.

The filter bed is also a good bacterial filter. The treated effluent can be disposed of by soil percolation or surface irrigation (Bowman and Bishaw 1991). The system is designed to serve 4-6 people for approximately 20 years after which the filter bed needs to be replaced. It is possible to increase the capacity of the system to serve more people or for an extended life span. This is a passive system with no requirement for power, chemicals or periodical servicing, except for the normal desludging requirements of the septic tank.

12.4.3 Sand Filtration

The Envirotech system (Figure 12.3) consists of a receival tank where settling of solids occurs, and a second chamber into which the effluent flows. When this is full, effluent is pumped to the top of a deep-bed plastic-lined sand filter. Effluent filters to the bottom of this device under gravity and flows back to a third chamber of the tank, from where the treated effluent is pumped to the irrigation field. General practice is to chlorinate in this final chamber, although it may not be necessary for subsurface irrigation. Systems are being installed in New South Wales and Queensland in Australia. Figure 12.3 is for a house with a family of five people.

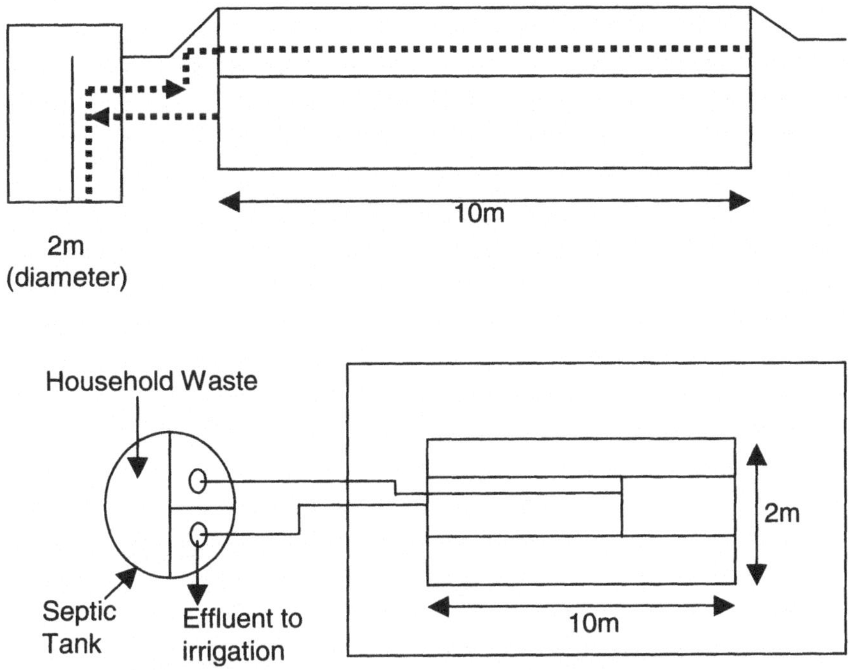

Figure 12.3 Schematic diagram of Envirotech Sand Filtration Unit.

12.4.4 Wet Composting

The Dowmus vermicomposting toilet system (Figure 12.4) can be upgraded to receive wastewaters - both blackwater and greywater (Cameron, 1994). In Canberra, (Australian Capital Territory) about 12 households have had trial systems installed for monitoring by Australian Capital Territory Electricity and Water (ACTEW) (Anon, 1996). Blackwater from the toilet enters a wet composting Dowmus tank and from there effluent goes to a second tank where greywater is also received. In this tank effluents are aerated around submerged volcanic rock media to achieve secondary standard treated effluent. From there the effluent goes to an irrigation storage tank in which chlorination occurs. The final effluent is mixed with rainwater to achieve further dilution and to improve the quality of water. The final solids product is vermicompost which is removed by an auger. This is an excellent soil conditioner.

12.4.5 Constructed Wetlands

Constructed wetlands (Figure 12.5) are based on natural wetlands. They act as a sedimentation, filtration and bioreactor 'vessel'. The wetland plants such as reeds are planted in a gravel bed. Wastewater effluent from septic tank is allowed to flow through the gravel system. The flow can be subsurface or surface. The organisms attached to the root system and the gravel degrade the organic matter. The plants take up nutrients and hence the wetland system is more effective in removing the nutrients than the conventional wastewater treatment system.

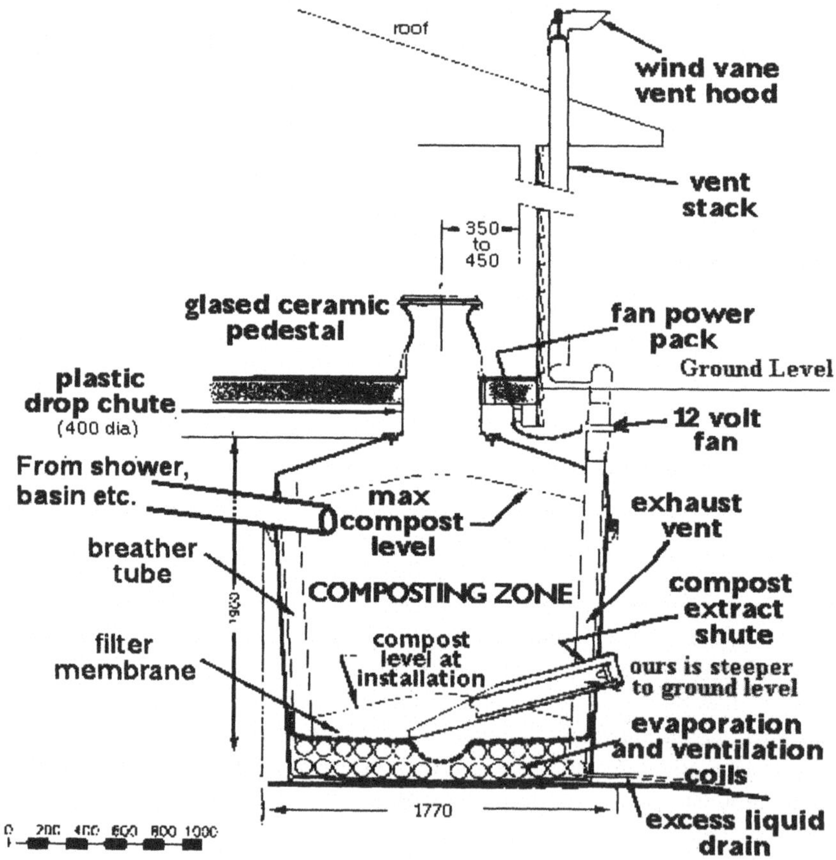

Figure 12.4 Dowmus Vermicomposting System (Adopted from http://www.hahaha.com.au/rammed.earth/toilet.htm 12/04/03).

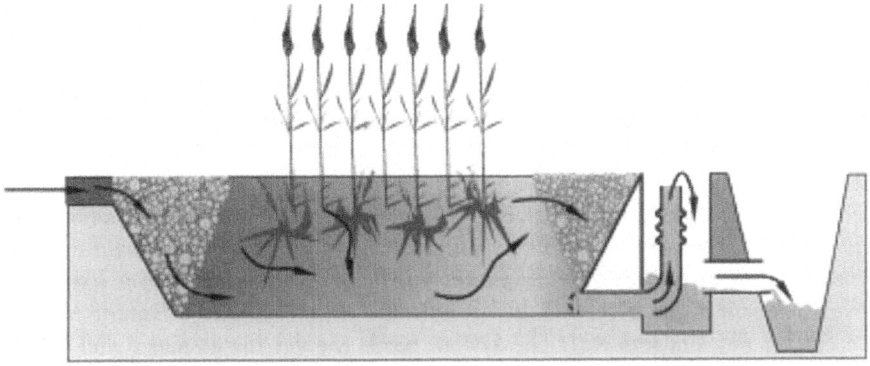

Figure 12.5 Constructed Wetland (Adopted from UNEP 2002).

12.4.6 Modified Aerobic Treatment Unit

The Biomax aerobic treatment unit (Figure 12.6) can be used for domestic wastewater reuse. Additional baffles have been installed in the anaerobic and aerobic chambers to enable more effective treatment of the lower BOD effluent input (c.f. combined blackwater and greywater). Effluent is irrigated to yards via 'Dripmaster' subsurface tubing.

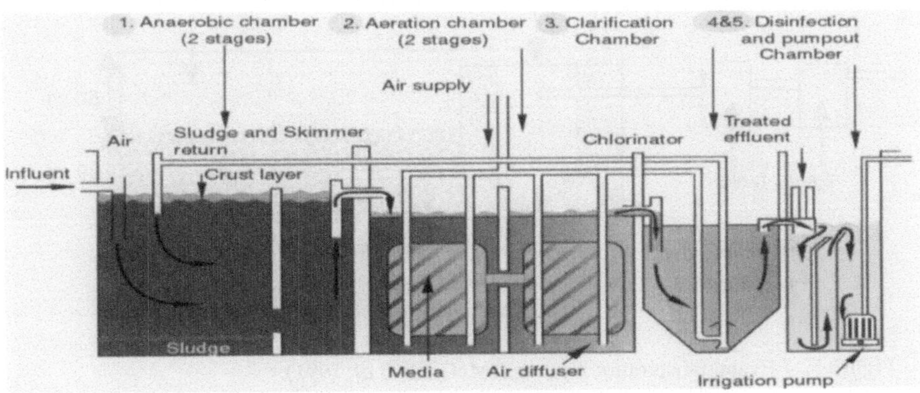

Figure 12.6 Aerated treatment unit (Biomax) (Adopted from UNEP 2000).

Biomax is an aerobic treatment system which provides a secondary level of treatment. This is available in two sizes. The domestic model is designed for 10 people and the commercial model is for offices, restaurants or other public institutions. The treatment system consists of a circular tank which is divided into four compartments (1) primary settling and anaerobic digestion tank, (2) aerobic chambers with fixed media and bubble aeration facility, (3) secondary sedimentation tank, the settle sludge is pumped back to the septic chamber, (4) chlorination and storage chambers. Chlorination is by tablet chlorinator and the final effluent is pumped for irrigation when the volume reaches a particular level. The soil at the irrigation area is amended with neutralised bauxite residue for further nutrient removal. The system needs regular maintenance and power requirements for aeration and pumping. The power requirement of 10 people/unit is 100W (2.5kWh per day).

12.4.7 Evapotranspiration Systems

Evapotranspiration (ET) systems (Figure 12.7) can be used in those areas where soil is comprised of more silts and clays and absorption fields have failed. These systems cost considerably less and require less maintenance than reticulated systems with lagoons (McGrath *et al.*, 1991). Effluent disposal in the ET trench occurs primarily by soil evaporation and plant transpiration rather than soil percolation - as occurs in conventional leach drains. The trench essentially comprises a layer of gravel for distribution of effluent below a layer of river sand through which capillary action to the surface occurred and in which plants grew.

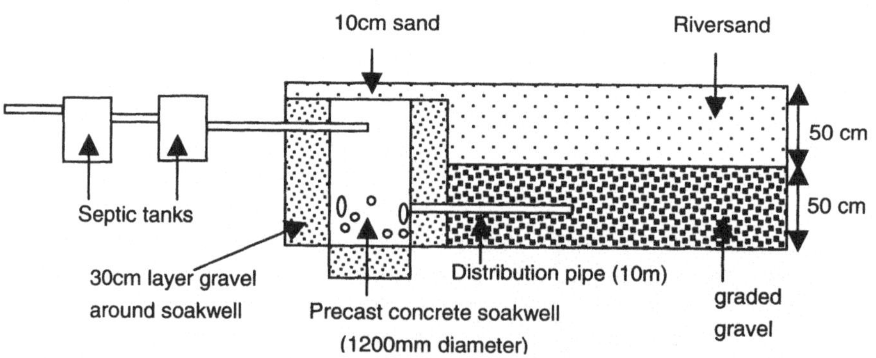

Figure 12.7 Evapotranspiration Systems (McGrath *et al.* 1991).

12.5 PROPOSED SYSTEMS FOR EACH SOIL TYPE

For cost-effectiveness and to most readily enable effluent reuse the following systems are recommended with respect to soil types:

Sand:	*modified aerobic treatment units; amended soil filters; sand filters; constructed wetlands* (avoidance of groundwater pollution);
Clay:	*evapotranspiration trenches* (avoidance of ponding); and
Rocky slopes:	*inverted evapotranspiration systems, sand filters* (avoidance of run-off).

The key problem to be overcome in each case is indicated in brackets. Wet or dry composting toilets can be used in conjunction with any of the above.

12.6 HOLISTIC DESIGN

Many concerns have been raised in relation to widespread implementation of greywater reuse without proper management or maintenance: reduced sewer flows, higher concentrations at treatment plants, public health risks, groundwater contamination, mosquito breeding in constructed wetlands, flooding during winter rainfall, sludge build-up and blockages. However, there is another issue for concern that may lead to some of these problems and others indirectly: poor design (or no design). Not just the design of the system itself but the manner by which the system is integrated into the landscape. Australian standards such as AS 1547 do, for example, specify minimum setbacks from houses and lot boundaries, provide ways of avoiding inundation and give design criteria for terraced disposal fields on slopes.

There are very few practical design methodologies that may serve the case of placement of a greywater recycling system in the house yard or community landscape. Two examples are:

* *hydroscaping* (Colwill, 1996) for sustainable garden aesthetic design; and
* *permaculture* (Mollison, 1988) for sustainable food production system design.

Hydrozoning will allow the placement of the greywater system in accordance with a garden layout designed for aesthetics and plant groupings of similar water needs.

The use of a design approach prior to installation enables placement of the greywater system in a landscape with respect to the vegetation type that it will support and its position in relation to other elements and natural influences on the site. If such considerations are ignored with a focus merely on the technical design of the system itself then improper management and maintenance and poor performance may still be the longer term outcome.

12.7 CONCLUSIONS

For the urban village, small country towns, or group housing a greywater reuse system utilising secondary treatment and disinfection maintained by a supplier may be most appropriate. For on-site greywater recycling at individual houses in a low-density settlement or remote community a primary system with large diameter subsurface irrigation 300 mm below the surface is appropriate using evapotranspiration in soils of low permeability. Filters, pumps and treatment units should be avoided as these may not be adequately maintained by the owner/occupier. Reuse from lagoons is commonly practiced in Western Australian country towns. If nutrient removal is necessary a treatment system such as Ecomax with sufficient vegetation to utilise the nutrient is ideal. Data-gathering on the long term effects of greywater on plants and soils and their nutrient uptake capacity is necessary. Field trials are necessary to optimise the irrigation fields for plant growth, particularly in the case of food species. Evapotranspiration systems, for example, have typically been designed too deep in the past for this purpose. A standard code of practice on greywater reuse should be adopted. If managed correctly wastewater reuse in remote communities can not only result in water savings but also improved public health through dust suppression from revegetation, improved nutrition from locally grown food, and less system failures from decreased loading on treatment systems. A summary table in which evaluations are made of some of the above technologies will be found in the first portion of Table 21.1.

12.8 REFERENCES

Anon (1996), Small-scale Systems Halve Water Use, *Waste Management and Environment*, **8**, 18 – 20

Beder S (1993), Pipelines and Paradigms: The development of sewerage engineering, *Australian Civil Engineering Transactions*, **CE35** (1), 79 - 85

Bowman M (1996), On-site Tertiary Treatment Using Ecomax Systems, *Desalination*, **106**, 305 - 310

Bowman M & Bishaw M (1991), *Ecomax Septic System, Explanatory Information – Ecomax Waste Management System*, Bishaw Forham Pty Ltd, Subiaco, Western Australia.

Cameron D (1994), Compost Filtration: A new approach to on-site resource management, in Mathew K & Ho G (eds), *Workshop on Localised Treatment and Recycling of Domestic Wastewater*, RADG, Murdoch University

Colwill J (1996), *The Waterwise Garden and How to Create One*, Water Authority of WA, Perth

Department of Primary Industries (1996), *Policy Options Paper - The Use of Greywater*, Brisbane

Fimmel B (1997), *personal communication*

Goodland R & Rockefeller A (1996), *What is Environmental Sustainability in Sanitation?*, Insight, UNEP-IETC Newsletter, Summer

Health Department of WA (1996), Draft Guidelines for Domestic Greywater Recycling, Water and Rivers Commission, Perth

Jeppeson B (1996), *Model Guidelines for Domestic Greywater Reuse for Australia*, Research Report # 107, Urban Water Research Association of Australia, Melbourne

Kinnaird M (1997), *pers. comm.*, Environmental Health Officer, Shire of Halls Creek

Kourik R (1995), Graywater for Residential Irrigation, *Landscape Architecture*, **85** (1), 30-33

Lugg R (1994), Health Controls on Wastewater Reuse, in Mathew K & Ho G (eds), *Workshop on Localised Treatment and Recycling of Domestic Wastewater*, RADG, Murdoch University, Perth

Mathew K & Ho G (1993), *Reuse of Wastewater at Jigalong Aboriginal Community*, RADG, Murdoch University

McGrath D, Ho G & Mathew K (1991), The Effectiveness of Evapotranspiration Systems in Disposing of Wastewater in Remote Aboriginal Communities, *Water Science and Technology*, 23, 1825-1833

Mollison B (1988), *Permaculture: A designer's manual*, Tagari Publications, Tyalgum

National Health & Medical Research Council (1996), *Draft Guidelines for Sewerage Systems - Use of Reclaimed Water*, National Water Quality Management Strategy, Melbourne

New South Wales Recycled Water Coordination Committee (1993), *New South Wales Guidelines for Urban and Residential Use of Reclaimed Water*, Public Works Department, Sydney

Newman P W G and Mouritz M (1996), Principles and Planning Opportunities for Community Scale Systems of Water and Waste Management, *Desalination*, **106**, 339 - 354

Niemczynowicz J (1993), Water Management and Ecotechnology - Towards 2020, *Workshop on Sustainable Urban Water Systems and Ecotechnology*, November 18, ISTP/RADG, Murdoch University, Perth

Standards Australia/NZ (1996), *On-site Domestic Wastewater Management*, AS/NZS 1547-Draft, Homebush/Well

Stone R (1996), Water Efficiency Program for Perth, *Desalination*, **106**, 377 - 390

UNEP (2002), *International Environmental Technology Centre Technical Publication Series No. 15*, United Nations Environment Program, IWA Publishing, London

Water Authority of Western Australia (1994), *Wastewater 2040 Discussion Paper*, WAWA, Perth

13

Arena: Municipal and non-aquatic recreational water use

13.1 INTRODUCTION

This chapter focuses on using water to maintain more extensive lawns and gardens than were the subject of Chapter 11, or on lawns with specialised usage. We will refer to the use of water in buildings associated with the lawns when extra factors are present over and above those discussed in the chapter 'Arena: Commercial water use inside buildings'. We have in mind sporting stadiums, municipal parks and sporting grounds, golf courses, gardens and nature strips between dual carriageways to give some examples. Market gardens and agriculture are not included. The use of irrigation for extensive lawns is a major issue, which will therefore be considered at a greater level of sophistication than in the previous chapter on 'External commercial water use'. Even so there will be material in this chapter that will be valuable in the arena of the previous chapter and vice versa. For example, the principles of hydrozoning apply equally well in the arena of this chapter, for high use areas such as cricket wickets and putting greens will need higher watering rates than will other turf. Our use of the word 'non-aquatic' in the title of this chapter flags that a separate chapter will be devoted to swimming pools or recreational aquatic centres.

13.2 CLASSICAL (SPRINKLER) TURF IRRIGATION WATER MANAGEMENT – A BRIEF OUTLINE

We break classical turf irrigation water management into two parts: an 'irrigation audit' and 'irrigation management'. We will treat each in turn below.

13.2.1 Irrigation audits

13.2.1.1 Sites

We assume that there can be multiple site usage; a typical example is where one local government controls multiple recreational park sites. Some of these recreational parks will have greater potential for water savings than others do. Where a single stadium or ground is to be audited, selection of some sites as having greater potential for water savings than other sites obviously becomes redundant.

Within the limited geographical range of a local government area we might assume that the climate is uniform, and that weather data representative of the whole could be applied throughout the area. Sometimes this will indeed be so. Yet some local government areas might include dramatic changes in relief. In such cases very localised weather data might be necessary. At the other extreme in Western Australia (for example) some local government areas, with minor variations in relief, cover an area greater than that of England. In such cases local weather data will need to be obtained corresponding to the actual sites. The data required are rainfall and pan evaporation, preferably with some statistics giving standard deviation within each month, or historical records of extremes.

At each site the historical records of water use are required. Bills from the water service provider will probably contain this information. Be aware that there might be more than one water meter at any one site. When historical water bills are not available or do not contain the information required, records of water meter readings will need to be taken over at least that portion of the year when the irrigation system is used. Where buildings are present on the site, the water use for them is required too. Sometimes this can be difficult to disentangle from the site irrigation water use. An estimate can sometimes be obtained of the non-irrigation water use by using the water service providers' meter readings in a wet portion of the year when irrigation water use is zero.

Later we will calculate maximum applied water allowance; this will require knowledge of the area irrigated at each of the sites to be audited. While the client or auditee may be able to provide this information, recourse might be needed to title deeds, maps, or earlier drawings using a planimeter to find the area. As a final resort, a surveyor can measure the area in the field.

The California State assembly in the USA has passed legislation (California AB 325) limiting the water that can be used in a project for landscape purposes. This requirement has been replicated in some other parts of the world, while in still others it is used as an informal, but accepted, rule of thumb for placing limits on the reasonable upper limit for landscape water use on a site. The formula (see Hunter, 1996) is presented below in a form modified for metric units:

$$\dot{Q}_{\max} = 0.8 A_l E_{Apan} \text{ L/day}, \qquad (13.1)$$

where $\dot{Q}_{\max}$ is the maximum applied water allowance in L/day

A_l is the area to be landscaped in m^2

E_{Apan} is the potential evapotranspiration in mm/day for the worst case

and 0.8 is a coefficient contained in the Californian legislation.

The maximum applied water allowance is calculated for each potential site the auditor might audit. These results are compared with the actual water usages and the site with the most potential for decreased water use (and thus savings in costs) is listed first, with others in order. Auditing will be undertaken in the order of the priorities established in this list.

13.2.1.2 Tuning

There is little point in proceeding with an audit of the sites selected if there are relatively small adjustments and repairs necessary to bring the system to a performance approximating its design specifications. Such a process of inspection, minor adjustments and repairs is called a system tune. The inspection will reveal the general condition of turf and plants, the existence of dry spots, the extent to which hydrozoning has been implemented, information about presence or absence of wind-breaks and other modifiers of micro-climate. Broken sprinkler heads will be replaced and defective alignment corrected. Systematic spatial degradations of turf or plant appearance can indicate the need for system pressure balancing or imply the location of obstructions in the distribution manifolds or lines.

13.2.1.3 Testing

The purpose of testing the system is to find out the performance of the irrigation system and to compare it with industry criteria. The method of testing the system is to use catch cups, set out in predetermined patterns between sprinklers, to find the application rate of water and other statistics discussed later. A catch cup is any circular shaped small vessel with vertical sides (or graduated non-vertical sides), which is sufficiently heavy to withstand wind gusts during the test. All catch cups in any one test should be identical. A typical (dense) layout of the catch cups is shown in figure 13.1. Note that the measurements are to be taken with typical water delivery pressures and with typical wind speed and direction for the site.

The rate of mean application of water is given simply by averaging water depths in the catch cups, ie:

$$\bar{\alpha} = \frac{60}{Tn}\sum_{i=1}^{n} d_i \qquad (13.2)$$

where	$\bar{\alpha}$	is the mean application rate in mm/hr,
	T	is the duration of the test in minutes,
	n	is the number of catch cups,
and	d_i	is the depth of water in the ith catch cup.

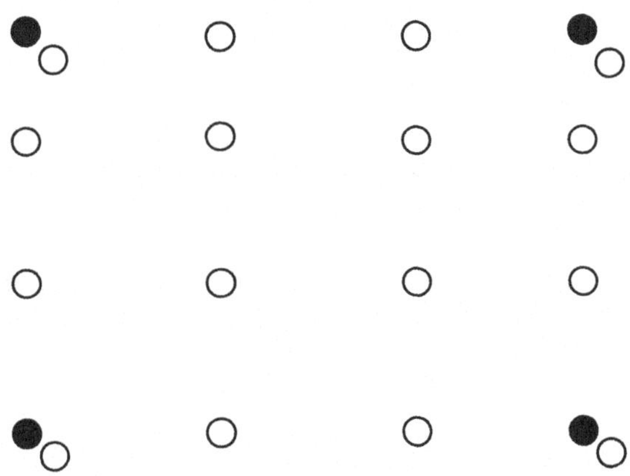

Figure 13.1. Sprinklers (●) with typical catch cup (○) arrangement – diagrammatic and not to scale.

'Coefficient of uniformity' (C_u) and 'Distribution uniformity' (D_u) are the two most common other statistics derived from the catch cup measurements. The first, named the Christiansen coefficient after its originator (Christiansen 1942) measures uniformity of application of water from an array of sprinklers, weighting equally sprinklers that over-water and under-water. It is given by:

$$C_u = 100\left(1 - \frac{1}{\overline{\alpha}n}\sum_{i=1}^{n}\left|d_i - \overline{\alpha}\right|\right) \tag{13.3}$$

An industry rule of thumb is to accept a system's performance when C_u is 85% or better.

Distribution uniformity is a more selective statistic, which emphasises the importance of the lowest 25% of catch cup readings. It is given by:

$$D_u = 100\left(\frac{\text{Mean of lowest 25\% of readings}}{\text{Mean of all readings}}\right) \tag{13.4}$$

Industry rules of thumb generally regard a distribution uniformity of 75% or greater as acceptable.

How are these statistics used? The mean application rate (13.2) is converted to a mean daily water usage and compared with the maximum applied water allowance (13.1). If it is greater, then the irrigation system is not operating properly and corrective action is needed. Yet the water auditor applies more stringent standards than this, for his/her goal is to conserve water, and reduce costs if possible, regardless of legislative or 'rule of thumb' requirements. Thus while irrigation operators might be satisfied with water usage below the maximum applied water allowance, the water auditor will examine possibilities for saving water even more carefully. The coefficient of uniformity, if unsatisfactory, tells the water auditor that sprinkler nozzles need examination and maybe replacement, that pressure uniformity in the system needs to be checked, that maybe there are blockages in the manifolds or distribution lines, or that windbreaks need to be considered. The 'distribution uniformity' coefficient is sometimes used to modify scheduling times, which we will consider now.

13.2.1.4 Datum schedule

The scheduling calculations in this chapter are presented without derivation. The datum irrigation schedule for the turf (or other plants) is the schedule required at the time of the year when the maximum irrigation water is needed by the plants. Calculating this schedule requires determination of seven parameters prior to undertaking the scheduling calculations.

First, the total available water capacity (W_{tot} mm/m) is necessary to find. An estimate can be obtained from Table 13.1. There is a fair spread to values tabulated in the literature and the tabulated values are indicative only. Accurate values for any given soil require direct measurement of that soil's moisture holding capacity.

Table 13.1. Indicative total available moisture in mm of water per 100cm of soil depth. Adapted from Brown (2001)

Soil type	Available moisture (mm water/m soil depth)
Sand	30 – 80
Sandy loam	70- 100
Loam	100- 150
Silt loam	150 – 180
Clay loam	140 – 160
Clay	140 – 150

Second, the design root zone (δ_{root}) is the depth of soil into which the main feeder roots of the plant extend. Typically for garden flowers and vegetables the depth will be 100 to 200 mm. For lawns 500 mm is a reasonable estimate, while many trees might be 500 mm to 1m or more. Citrus trees are surface feeders and 100 mm is a reasonable value for them. For any particular plant type, δ_{root} can be found by trenching and observing the root depth.

Third, the management allowable deficit (μ_{ad}) is required. Generally with a value between 0.4 and 0.6, this factor represents the proportion of the total available water capacity that we allow to be drawn down and thus which requires topping up.

Fourth, The plant crop factor C_c is required. C_c is that proportion of the potential evapotranspiration rate required by the plant to be healthy. The crop factor should be evaluated locally in the appropriate soil and climatic conditions. Often such data is not readily available, and we recommend a value of 0.6 as a starting point for warm season grasses (Short & Colmer, 1999a and Short & Colmer 1999b).

Fifth, the potential evapotranspiration can be estimated by several methods, each with strengths and weaknesses. Here we choose to use the evaporation from an A-type pan (A-pan) as a straightforward, accessible and useful method, with extensive data likely to be available from local weather stations. Where such data is not available, the measurement of pan evaporation is relatively simple compared with other methods. As a rough guide we present indicative values in Table 13.2, deriving from Brown (2001). Average daily A-pan evaporation (E_{Apan} mm/day) for the month of highest evaporation are tabulated in Table 13.2 for a variety of climatic types. Where possible, local values should be obtained from the local weather stations.

Table 13.2. Indicative potential evapotranspiration rates.

Climate category	**Average midsummer maximum temperatures Deg C.**	**Average midsummer relative humidity %**	**Potential evapotranspiration mm/day**
Cool humid	<21	>50%	2.5 to 3.8
Cool dry	<21	<50%	3.8 to 5
Warm humid	21 to 32	>50%	3.8 to 5
Warm dry	21 to 32	<50%	5 to 6.3
Hot humid	>32	>50%	5 to 7.6
Hot dry	>32	<50%	7.6 to 10

Sixth, the application rate ($\dot{Q}$ mm/hr) that the soil will absorb without runoff is necessary. Data to enable such a calculation to be carried out are tabulated in Hunter (1996), deriving from the Department of Agriculture, USA. Here we translate the table into Metric units, and present the results in Table 13.3

Seventh and finally, the irrigation efficiency (η_{irrig}) is required. This is the amount of water the plants receive (as measured by catch cups for example) compared with the amount of water delivered to the irrigation pipe distribution system. This should be measured.

Table 13.3. Maximum application rates in units of mm/hr for a variety of soils to avoid surface runoff, adapted from Hunter (1996). G: groundcover. N: no groundcover

Soil type	**Slope 0%-5%**		**Slope 5%-8%**		**Slope 8%-12%**		**Slope >12%**	
	G	**N**	**G**	**N**	**G**	**N**	**G**	**N**
Course sandy	51	51	51	38	38	25	25	13
Course sandy over compact subsoil	44	38	32	25	25	19	19	10
Light uniform sandy loam	44	25	32	20	25	15	19	10
Light sandy loam over compact subsoil	32	19	25	13r	19	10	13	8
Uniform silt loam	25	10	20	10	15	8	10	5
Silt loam over compact subsoil	15	8	13	6	10	4	8	2.5
Clay or clay loam	5	4	4	2.5	3	2	2.5	1.5

We now turn to the scheduling calculation proper. The net water requirement (Q_{net} mm) is given by:

$$Q_{net} = W_{tot}\delta_{root}\mu_{ad} \tag{13.5}$$

The gross water requirement (Q_{gross} mm) is:

$$Q_{gross} = Q_{net} / (\eta_{irrig} /100) \tag{13.6}$$

The duration of the irrigation (τ_{irrig} hrs) is:

$$\tau_{irrig} = Q_{gross} / \dot{Q} \tag{13.7}$$

and the interval between irrigation start times (T days) is

$$\mathrm{T} = Q_{net} / \left(E_{Apan} C_c\right) \tag{13.8}$$

Here we make the assumption that the soil already holds its total available water and that the irrigation tops up the available soil moisture as it is depleted.

Notice that in equation 13.6, we use the irrigation efficiency η_{irrig}, following Milani (2001). Some other authors use the coefficient of uniformity or the distribution uniformity of equations 13.3 and 13.4. As water auditors we believe this is unacceptable, for it can undermine the urgency of undertaking the irrigation audit portion of the process we have described, by simply increasing water usage to compensate for poor tuning and testing. Looked at another way, the larger portion of the turf is over-watered, so that a small portion receives sufficient. Thus improvement in the distribution uniformity is a crucial tool for overall reduction in water use. (Other scheduling coefficients are available in addition to the distribution uniformity. For further details see, eg., Johnston 1996).

This resultant irrigation schedule is called the datum (or base) irrigation schedule. Establishment of this schedule completes what is called the 'irrigation audit' in the industry. We move now to brief remarks on irrigation management.

13.2.2 Irrigation management

13.2.2.1 Datum schedule introduction and adjustment

Implementation of the datum irrigation schedule constitutes the first step of the irrigation management process. Yet this schedule is not appropriate for any conditions but the maximum irrigation conditions for which the irrigation is installed. Therefore the datum irrigation schedule is adjusted downwards in response to natural rainfall, higher humidity, lower temperatures, lower solar insolation or lower wind-speeds; in short the irrigation schedule is adjusted to accommodate lower levels of A-pan evaporation. Indeed local A-pan evaporation data can be used to reschedule the irrigation controller program. Equation (13.5) can be used to calculate the irrigation schedule in these less demanding circumstances, and the procedure following (13.5) above will issue in a new irrigation schedule. Even so, the practical experience of irrigation operators often is given great weight in determining precisely what the discounted irrigation schedule will be. There is no universal answer to the question of whether confidence is well placed in the experience of irrigation operators.

13.2.2.2 Actual water use

Keeping records of actual water use and comparing it with calculated water requirements, is a valuable way of detecting deviations of system performance from design criteria.

13.2.2.3 Maintenance

Maintenance of irrigation hardware is essential to minimise water use. Worn or broken sprinkler heads need to be replaced. In doing so, it is crucial that the performance specifications of the original sprinkler heads be replicated. Failure to do so will almost always lead to a decrease of distribution uniformity and thus to brown patches and so to an increase of water usage. Pressures throughout the system also need checking, for deviations from the original design pressures likewise will lead to erosion of the distribution uniformity and so to an increase of water usage.

13.2.3 Reflections on classical (turf) irrigation water management

Clearly the process of irrigation auditing and irrigation management is of concern to the water auditor, for both are accompanied by the potential to save significant quantities of water. Because of wind and evaporative losses, we might ask whether sprinkler irrigation is an irrigation method a water auditor might live with happily, or whether other options might be more attractive in many circumstances? We believe the answer to this question is 'yes'. Therefore, we will turn our discussion to two further topics: soil moisture sensors and below ground irrigation, in order to complete our discussion on turf irrigation.

13.3 SOIL MOISTURE SENSORS – IMPACT ON IRRIGATION MANAGEMENT

There is a wide variety of soil moisture measuring sensors available. Here we will focus on one type only, capacitive sensors that measure the dielectric value of the mixture of soil and water. The dielectric value of the mixture of soil and water depends strongly on the water content. Thus changes in dielectric value can be interpreted very accurately as changes in the volumetric soil moisture content of the soil (See Johnston, 1996). Soil moisture sensors can be stacked in a column, so that soil moisture content can be measured at a variety of depths. Typically the depths will be at 100 mm intervals to a depth of 500 mm. Underground cabling can convey signals from the sensors, to a central computer station, where the data can be processed. Alternatively radios can transmit the data to the central data logging and data processing computer. A profile of soil moisture to a depth of 500 mm is thus available at a series of locations where the stacks of probes have been installed. In order to make use of this data, we need to define two terms.

The *field capacity* or *total available water capacity* is the moisture content of the soil (mm) at which water moves beyond the root zone at a rate that is acceptably *low*. By contrast, the *refill point* is the moisture content of the soil (mm) at which the first signs of plant stress become visible. Ideally the water content of the soil falls between these two extremes, so that water is not wasted bypassing the root zone and so that the turf (or other plants) do not suffer from drought stress. Both the field capacity and the refill point are determined empirically in the field. Irrigation is initiated by the computer controller when the soil moisture level drops to the refill point and it is continued until the soil moisture content reaches the field capacity. Figure 13.2 exemplifies the soil moisture in the turf root zone using this procedure.

In field tests described in Johnston (1996) the managed irrigation system described above sampled soil moisture at 10 minute intervals and its performance was compared with a badly tuned manual system used over 500ha of turf. The managed irrigation system gave rise to reduction in water use of about 70% and a return on investment of approximately 400% per annum based on savings in energy use alone. Clearly such a system of capacitance soil moisture sensing and computer control has much to commend it, even if a comparison with a badly tuned manual system shows it in its best light. For example, the issue of water leaching past the root zone is dealt with by direct measurement, rather than by semi-empirical modelling. Irrigation scheduling ceases to be a difficult and continually changing task, as the instrumentation accomplishes the scheduling automatically. Modelling of the complex of variables, such as temperature, humidity, solar insolation, crop factor, management allowable deficit, is eliminated by the direct measurement of soil moisture. Problems of overuse of water due to low distribution uniformity are not overcome, however, even with the use of multiple ranks of soil moisture sensors located at a variety of sites. Distribution uniformity remains a problem associated with distribution piping and sprinklers themselves, rather than with the irrigation scheduling and management.

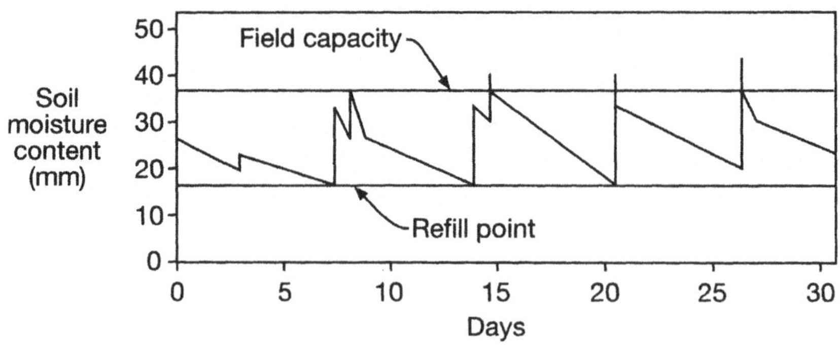

Figure 13.2. Typical soil moisture content under typical computer management using capacitive soil moisture sensor arrays. (Adapted from Johnston 1996)

The water auditor will only occasionally be in a position to determine whether capacitive soil moisture sensors and irrigation controllers are employed. When s/he is in such a position, we strongly recommend going down this path, for the savings in water and money can be very substantial indeed, despite the initial capital cost.

There are intrinsic difficulties with above ground sprinkler systems. We summarise them here: increase in fungal diseases, wind drift, vandalism, evaporative losses, loss of turf facility while watering is in progress, restricted optimal watering time (ie early morning). We acknowledge the positive aspects of sprinkler systems too: cheaper than below ground drip irrigation, usually easy to maintain and to operate, and versatile in the situations to which it can be applied. Even so drip irrigation, which can be applied either above ground or below ground, has some compellingly positive features.

13.4 DRIP IRRIGATION

Drip irrigation was invented in the 1960s. Since then it has gained wide acceptance in agriculture, while gaining increased recognition as an option in landscaping. Vandalism has become an increasing problem in countries with 'developed' economies. Drip irrigation mitigates this problem, especially in its below ground form and this has accelerated its use in landscape irrigation. Because drip irrigation applies water close to the root zone of plants and at low application rates, run-off and leaching past the root-zone reduces as does the likelihood of consequent problems with algal blooms in nearby surface expressions of the water table. In particular these two issues have aided the acceptance of drip irrigation, yet there are further advantages too.

In its above-ground form drip irrigation is cheap and easy to install, reduces weed growth, makes disease control more easy, enables flexible watering times, applies water uniformly, and it can accommodate odd shapes and narrow strips. Drip irrigation admits the use of some-what more saline water than sprinklers do and applies water at a low rate, with implications for smaller pipe sizes and costs. Alongside these advantages we recognise that not all situations are best served by drip irrigation. The quality of the water is important, as drippers can clog with depositions more readily than sprinklers do. Thus filters and their selection are important and in some instances chemical water treatment might be warranted. On top of this, the irrigation design must be thorough, as short cuts can result in very poor performance.

When drip irrigation is applied below ground, the advantages listed above continue to apply. Some are strengthened. For example the chance of vandalism is reduced further. Of course costs rise as we move from above ground to below ground irrigation, as does the difficulty of installing it.

From the water auditor's point of view the use of below ground or below mulch drip irrigation reduces evaporative water losses, results in a very high distribution uniformity and is not accompanied by windage losses to which sprinklers are prone. Thus when a water auditor has the opportunity to recommend below mulch or below ground drip irrigation, we strongly recommend that s/he does so. As outlined in the chapter 'Water management strategies', an economic evaluation must be made. We believe such an evaluation will often prove positive.

Soil moisture sensors and computer controlled irrigation schedules that start irrigating when the soil moisture has dropped to the refill point, and stop when the field capacity is reached offer great potential water savings. These savings are potentially higher still when the system is used in conjunction with below mulch or below ground drip irrigation systems. Not only is the efficiency of water use increased markedly, but also the quality of groundwater (and its surface expressions) is protected too. Thus the cause of water conservation is broadly fostered by the adoption of these strategies. Very often they will also prove very economically attractive.

13.5 STADIUMS AND BUILDINGS IN SPORTING GROUNDS

13.5.1 Introducton

The major focus of this chapter, that is turf for sporting and other municipal-park use, cannot be taken in isolation from its context. The context is often a stadium, or other less substantial buildings, which nevertheless form a part of the total human purpose for the site. There are some aspects of the use of water in such buildings, which deserve comment that complements the chapter 'Commercial water use inside buildings'. We now turn our attention to those aspects.

As a generalisation, the major area of water use in buildings will be the in the bar area in a sporting ground. Thus this area should be the focus of substantial efforts to reduce water usage or to engage in grey-water re-use. Bear in mind that the requirements of local government and health considerations can impact upon whether, or how, grey-water recycling might be implemented. Again in general terms, it should be applied below ground. Automatic flushing of grey-water lines to prevent build up of deposits will usually save water, and employees' time, compared with manual flushing.

During breaks in play, toilets will receive particularly heavy use, showers will receive heavy after-game use, while kitchen and dishwasher use will be significant, though less than the former. We expect these areas to yield savings in water use under the scrutiny of the water auditor, using the same techniques as discussed under 'Internal commercial water use'. Additionally we note that replacing old shower-heads with water efficient devices can sometimes lead to complaints about the difficulty of washing hair. Hand held shower-heads can reduce this problem in areas where the demand for hair washing is high. Foot controlled taps and spring loaded taps are useful for reducing water use in the kitchen or canteen areas.

13.5.2 Overall site issues

Opportunities are available for saving on the installation costs of piping and on the overall energy cost of the site. This can be achieved by scheduling some water-use, and thus energy use, activities so that they do not correspond with the peak of either the daily water or energy use curve or the seasonal water or energy use curve. In order to make such decisions data needs to be acquired to enable construction of daily and annual water and energy use curves, ie time series. We raise two examples. (i) Typically the optimal time for watering turf is in the early morning, prior to the 'breakfast spike' in the power load curve. Often advantage can be taken of lower electricity tariffs by watering at that time. Against this proposal might be the need to spread the watering over a longer period of time to maintain water flow rates within the capacity of installed pumps and pipes. (ii) An activity like flushing grey-water distribution pipes could be timed to occur when electricity tariffs are lowest.

13.6 REFERENCES

Brown, F. 2001 *Personal communication.* Frank Brown irrigation consultants.

Christiansen, J.E. 1942 *Irrigation by sprinkling.* University of California, Agric. Exptl. Station, Bulletin 670. 124pp.

Hunter Industries Incorporated 1996. *The handbook of technical irrigation information.* Hunter Industries Incorporated, San Marcos, USA.

Johnston,K.J. 1996 *Turf irrigation and nutrient study – turf manual.* Royal Australian Institute of Parks and Recreation – W.A. Region, Perth, Western Australia.

Milani, S. 2001 *Applied landscape irrigation design course materials.* Environmental Technology Centre, Murdoch University, Perth, Western Australia.

Short, D. & Colmer, T. 1999a. An evaluation of the water requirements for a diverse range of turf species under WA conditions. *Parks and Leisure,* **2(1),** 8-12.

Short,D. & Colmer, T 1999b. A comparison of turf grass genotypes during summer in Perth. *Australian Turfgrass Management,* **1(5)**, 20-22.

14

Arena: Aquatic centres

14.1 INTRODUCTION

Swimming pools or aquatic centres have become commonplace in nations with advanced economies. Large cities or towns in countries with less developed economies will often have aquatic centres too. By aquatic centres we mean housed complexes of swimming, wading and diving pools, sometimes both indoor and outdoor within the same complex. Often cafeterias, and other shops, are located within the centre, along with the usual toilet and shower facilities. Spas frequently will be present too, and sometimes saunas. Such centres are large consumers of water and therefore attract the interest of the water auditor. As indoor commercial water use and landscaping water use is treated earlier in this book, we will focus now on the pools and associated machinery for cleaning and maintaining the quantity and quality of water in the pools. Thus at the end of this chapter you should be able: to list the major sources of water loss, to understand the role of evaporative water losses, to understand the reasons for filter backwash and typical flows and to discuss a range of other issues including environmental conditions. Evaporation, filter backwash and spa activities are the main water consumers. In the following sections we will consider major processes by which water is lost and some other matters of use to the water auditor.

14.2 EVAPORATION

Evaporation is the single most significant process by which water is lost from swimming pools. It is significant whether we are referring to inside or to outside pools. The water auditor will need to estimate this loss from the swimming pool in order to obtain closure of the water balance. There are three obvious ways to achieve this.

First, in principle the pool could be operated without bathers and water treatment for a period during which no water was added or removed. The evaporation could be measured directly as the drop in pool level over a period of time (at least one day), assuming no leakage from the pool. This will often be impractical. Most operators of public pools would strongly resist excluding swimmers from the pool for such a period. Assuming that there are no leaks in the pool is usually a spurious assumption. Many pools leak; a leakage as large as 5% of pool volume daily is not uncommon, while some leak even more. Thus we need to look for other ways to measure or to estimate evaporative water losses.

A second way could be to measure A-pan evaporation at the site of the pool. This method is simple and straightforward, but is not foolproof.

A third way is to calculate the evaporation using one of many available formulae. It is beyond the scope of this book to evaluate a particular approach. (Singh & Xu 1997 do this and show there might be little to choose between competing methods.) We simply choose an approach that has a solid physical foundation. It includes the heuristically important variables: vapour pressure at the temperature of air well above the water, saturated vapour pressure at the water surface temperature and wind speed at say 2m above the water surface. (Even so Singh & Xu 1997 suggest that the wind speed term in the equation we choose might have little effect). We choose the equation of Penman (1948), reproduced below with the constants adjusted, for we have moved away from Penman's use of Imperial Units.

$$E = 0.262(1 + 0.146u_2)(e_{si} - e_a), \quad (14.1)$$

where E is the evaporation rate of water in mm/day,
u_2 is the wind speed 2 m above the pool,
e_{si} is the saturated vapour pressure at the air water interface at the temperature of the water surface,
and e_a is the vapour pressure of the air well above the water surface (say 2m).

In order to calculate E values need to be determined for, e_{si}, e_a and u_2. Of these u_2 is most easily determined, with a hand-held turbine velocity meter for example. The vapour pressure values require a little more calculation. Starting with e_a, we note that the ambient air water vapour pressure is identical to the saturated vapour pressure at the dew point. Thus given the air dry and wet bulb temperatures, the dew point temperature T_{da} can be determined from a psychrometric chart. We repeat that the saturated vapour pressure at T_{da} is identical to the vapour pressure of water in the ambient air. The saturated vapour pressure of water can be determined from an approximation to the Clausius-Clapeyron equation, given in Linacre & Geerts (1997):

$$e_s = \exp[18.956 - 4030.18/(T - 38.15)] \quad (14.2)$$

where e_s is the saturated vapour pressure at the temperature T in degrees Kelvin.

Substitution of T_{da} into equation (14.2) gives the vapour pressure of the ambient air above the pool, e_a. Substitution of the surface water temperature (in degrees Kelvin and preferably measured with an infra-red thermometer) into (14.2) will also give e_{si}. The evaporation rate of water from the swimming pool can now be calculated from equation (14.1). To get an average value over a day, ambient wet and dry bulb temperatures and surface water temperatures need to be taken several times over the course of the day at equal intervals. E is calculated for each sample and averaged over the day.

Some cautionary remarks are in order. Singh & Xu (1997) show compellingly that the constants in equation (14.1) are specific to the particular water body to which the equation is being applied. Thus if accurate results are required, the constants in (14.1) should be calibrated at the swimming pool. In still conditions the leading constant can be determined and then the constant associated with u_2 can be found with a fan at a variety of speeds blowing across the surface of the water in the container. Such a calibration process is not trivial.

Outdoor pool complexes usually include a shallow wading pool, where evaporative losses may be significant, as well as diving pools and competitive pools in standard 25m and 50m lengths. In outside (non-wading) pools, the water temperature rarely climbs above 23-24°C in mid latitudes. In indoor centres, the pool water temperature may be as high as 30°C. We might anticipate higher evaporation in the indoor pool due the higher water temperature, yet this will often be substantially offset by windier conditions over an outside pool.

Determination of evaporative water losses from the pool might be tedious, but without such an estimate no assessment can be made of leaks from the pool from a water volume balance (closure).

14.3 TYPICAL POOL WATER TREATMENT CIRCUIT

We will briefly outline a typical pool water treatment circuit as represented in Figure 14.1, so that the reader understands the major processes and components, before we discuss the typical second major water loss – filter backwash. The balance tank at the bottom of the figure enables the achievement of relatively stable water levels in the pool itself. Overflow water from the pool spills into the balance tank. In the tank various water treatment chemicals are added to maintain water quality. Treated water is then drawn from the balance tank to be pumped to the pool.

Three pumps are shown in Figure 14.1, of which two will be used while the other is on stand-bye. The stand-by pump makes it possible for the pool water treatment system to operate normally while servicing glands or while major repairs are conducted on the third pump. From the pumps the water is passed (top of the figure) through the three sand filters and then the filtered water (bottom of figure) is piped to the pool. In time the sand filters gather overmuch filtered material to continue operating acceptably. This material is removed from the filters by two means: first, air under pressure is forced through the filter in the reverse flow direction (air scour), breaking up crusts of fat that might trap other materials held by the filter; second, the water flow is reversed through the sand filters by operating the valves labelled in Figure 14.1. The operation of passing water in the reverse direction through the filter is called backwashing.

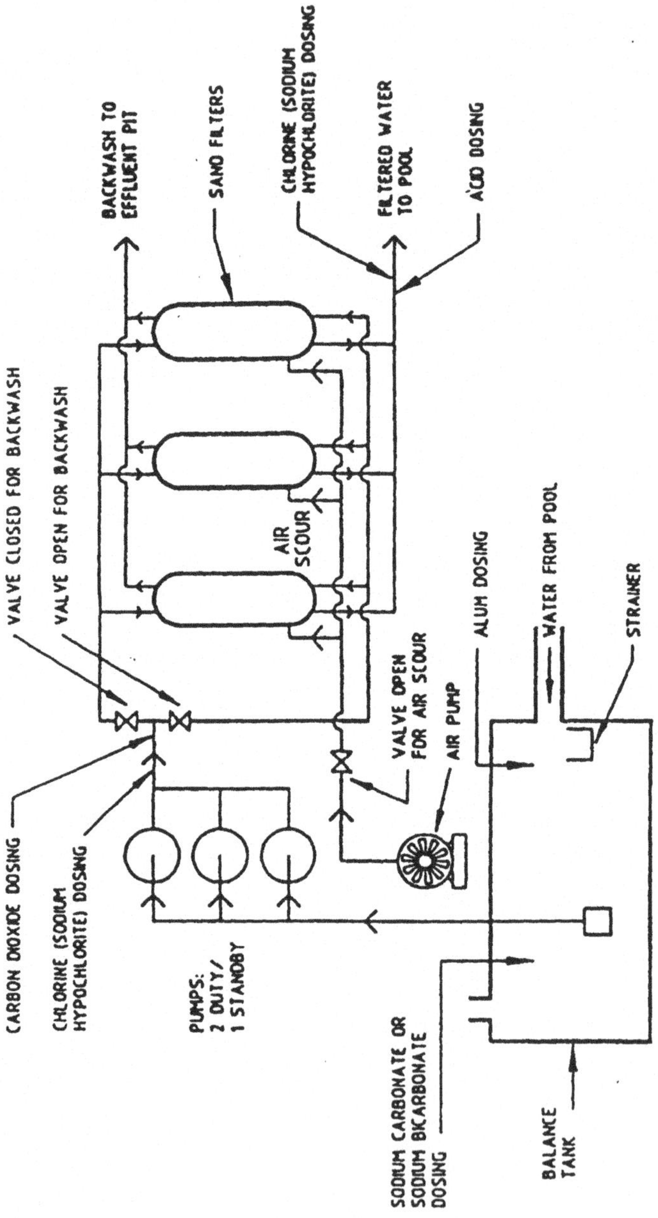

Figure 14.1 Water treatment circuit for a typical pool. Note that the pool itself is not represented in this figure. *Source: John Schlafrig and Mel Rowe.*

14.4 FILTER BACKWASHING

Filter backwashing is necessary for two reasons, of which we introduced the first above - to clean the filter. The second reason is that at very high bather load, it can be necessary to exchange the water more quickly than otherwise, in order to control water quality. Thus at high bather load backwashing is necessary daily, for example in mid-latitudes at mid-summer, when perhaps five thousand people per hour can pass through a pool. After evaporation, filter backwash is likely to be the greatest loss of water in the aquatic centre.

We present now some useful indicative values for the water auditor. For a *typical* indoor centre:

- pool volume ≈ 1,700 cubic metres
- pool turnover ≈ 4 hours or less (down to 2 hours)
- flow per hour ≈ 425m^3 or more
- number of filters 3
- flow per filter ≈ 142 m^3/hour
- backwash per filter ≈ 30 m^3
- total backwash ≈ 90m^3/backwash
- % make up water needed ≈ 5 per cent

The backwashed water is usually discharged, often to a ground-water recharge pit or to a stormwater drain. It could, however, be reused for a beneficial use, eg. for landscaping irrigation. The water auditor actively seeks to find such beneficial uses for reject-water.

14.5 POOL PIPEWORK LEAKS

As noted earlier, an unsuccessful attempted water volume balance (ie. attempted closure) including calculation of evaporation, can reveal leaks from either the pipework associated with the pool or with the pool itself. Leakage from pools is not unusual and it is often substantial.

14.6 SPA SYSTEM

Figure 14.2 shows the layout of a typical spa and sauna. This system is usually self-contained complete with a sand filter which needs periodic backwashing. There is usually significant evaporation as a spa operates at 40°C. Water losses can also occur in the spa because of spillage and displacement. Checks should be made of the number of occupants in the spa at peak times as overloading results in spillage by displacement. The filter associated with the spa would typically be backwashed daily, losing 15% to 20% of the spa's volume. A representative volume of backwash water would be 2 m^3 daily.

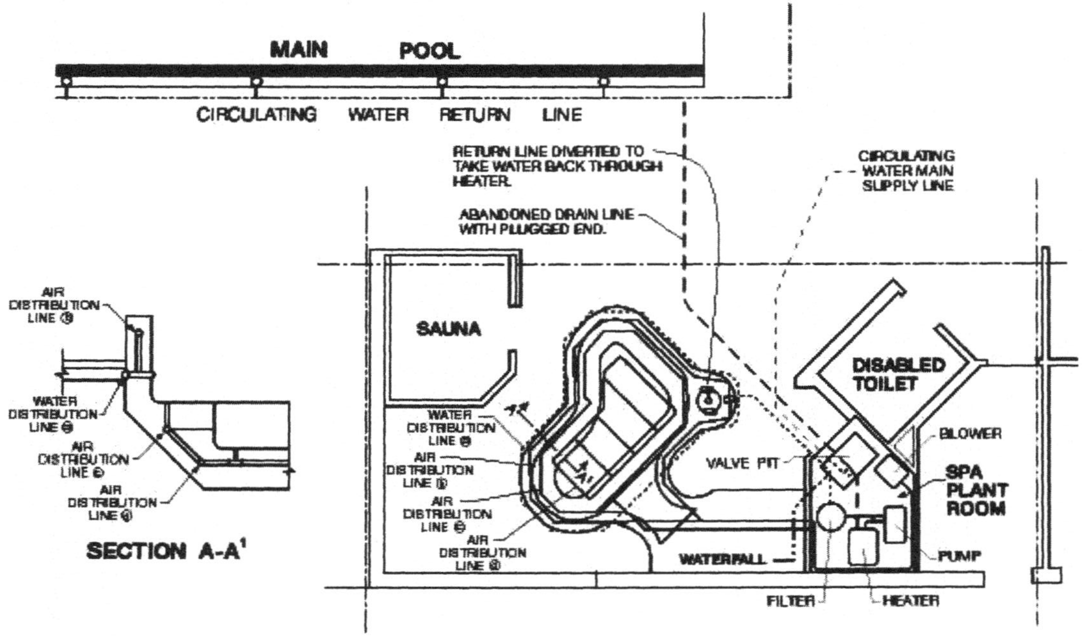

Figure 14.2 Typical layout of a spa and sauna. *Source: John Schlafrig and Mel Rowe.*

14.7 SUNDRY LOSSES AND BOILER BLOWDOWN

The water auditor should examine the plant-room carefully for water losses, especially from hoses and glands. Both chemical treatment and blowdown will control scaling and salinity in the boiler. The water auditor carefully records the frequency and volume of blowdown and encourages the centre managers to optimise the process.

14.8 DOMESTIC WATER USE

Though internal commercial water use has been treated in an earlier chapter, we remind the reader that kitchen, shower, toilet and other domestic water use needs to be included in the total water audit of the aquatic centre.

14.9 ENVIRONMENTAL CONDITIONS

Waste energy in the aquatic complex can raise the energy use of air conditioning where installed, or contribute to lower humidity and thus higher evaporation losses. Thus we encourage careful attention to energy efficiency in the complex. Correctly adjusted ventilation will contribute to an acceptable level of humidity while minimising evaporative losses of pool water.

14.10 REFERENCES

Linacre, E. & Geerts, B. 1997 *Climates & weather explained.* Routledge, London. (Includes CD).

Penman, H.L. 1948 Natural evaporation from open water, bare soil and grass. *Proc. Roy. Soc.* A193, 120-148.

Singh, V.P. & Xu, C.-Y. 1997 Evaluation and generalization of 13 mass-transfer equations for determining free water evaporation. *Hydrological Processes,* **11**, 393-411.

Tamminen, T. 1996. *The pool maintenance manual.* McGraw-Hill, New York., 468 p.

15

Water conservation in isolated communities

15.1 INTRODUCTION

We intend this chapter to raise some of the issues in water conservation in both remote communities of indigenous peoples and in remote non-indigenous settlements. We present material that derives from our experience of remote Aboriginal communities in Australia, believing that the essential principles are common to a much wider range of particular communities world wide. We do caution that local environmental and cultural factors can strongly affect the appropriate implementation of water conservation measures in any particular community.

Indigenous Australians had an extremely sensitive water care ethic as part of their traditional nomadic lifestyle, particularly in the more arid regions (Flood, 1983). Their move to sedentary lifestyles occurred as the result of earlier forced relocation into missions, government settlements and pastoral stations. Today there are many discrete, remote, indigenous communities in the arid regions of Australia. For example, there were 367 communities recorded in Western

Australia (WA) in 1997 by the Federal Government's Aboriginal & Torres Strait Islander Commission, which included large settlements, town reserves, small outstations and seasonal camps.

The development of water supply and sanitation in these communities over the last 100 years has some similarities to developments over a much longer time period in the European cities of the Industrial Revolution, commencing some 200 years ago. Water supply without adequate quality, or wastewater treatment, characterised both cases, with appalling public health consequences. The last 10 years have seen considerable improvement to housing and infrastructure in many remote indigenous communities, and although gains have been made in public health standards, there is still much progress that needs to be made before mainstream health indicators are matched (Pholeros *et al.*, 1993).

The approach to improving the level of service in the remote indigenous communities has generally been to adopt the same technology-practice as that used in urban areas when the limited finance finally becomes available for such works. Equality of inputs, ie. the technological hardware for water and sewerage, has been understood by city-based engineers and bureaucrats as the means by which to achieve 'equity'. However, the result of this approach has not always been equality of outcomes - an acceptable quality of life, similar health indicators and the sustainable management of resources (Race Discrimination Commissioner, 1994). This technology transfer is not usually a simple process. During consultation, engineers and bureaucrats, without understanding the longer term impacts of technology choice, often respond directly to the imputed needs of community residents. These needs are often perceived by engineers to be the same as those for non-indigenous Australians in urban areas.

Consequently, hardware has often been delivered to the communities without the management and maintenance systems to support it and without institutional capacity building. The need for the latter has not been recognised within both indigenous and government organizations. This need was made quite clear in the recommendations of the 1991 Royal Commission into Aboriginal Deaths in Custody, but has not been adequately followed up by Government. Rather Governments have at times thwarted the formation of regional indigenous organizations. Such organizations were intended to support autonomous housing and infrastructure as a means of self-determination, training delivery and employment generation, while fostering appropriate technology-practice.

Issues of sustainability of water resources have generally been ignored in the development and expansion of remote indigenous communities in arid regions. However, as there are still many communities where considerable housing and infrastructure works need to be undertaken, one can argue there is greater scope for the ready introduction of holistic water conservation strategies than in the large Australian cities, where technology and lifestyle practices are well established.

15.2 BACKGROUND

The Remote Area Developments Group (RADG) of Murdoch University conducted research to identify the needs and opportunities for water conservation in remote Aboriginal communities, prepare promotional materials, including a booklet and video, and to establish some trials of new technologies.

Unlike many of the large urban areas of Australia, Aboriginal communities in arid lands do not have a diverse range of water sources. Typically, there are groundwater sources whose sustainability in the face of growing populations is uncertain. Yandeeyarra in the Pilbara and Jigalong in the Western Desert of Western Australia are examples of communities that may have limited opportunities for further social and economic development due to limited groundwater supplies. At Coonana, in the Western Australian Eastern Goldfields, where there is no groundwater, dams are used, harvesting water for general usage from very unpredictable rainfall. Water shortages have been extreme (Race Discrimination Commissioner, 1994). Even so there has always been drinking water available, due to the installation of large rainwater tanks at each house. With the current water supply schemes in these communities it is not possible to have some highly sought after recreational opportunities - for example, to have swimming pools or grassed football ovals.

Water conservation may not immediately realise these recreational opportunities but it may allow development in other areas such as future population growth, revegetation, small areas of lawn in house yards, and plant propagation nurseries. If water conservation strategies are adopted alongside community development, the necessary improvements to living conditions can occur while enhancing the sustainability of groundwater resources.

15.3 APPROACHES TO WATER CONSERVATION

Water conservation can be tackled from three angles:

1. lifestyle changes (e.g. take short showers, fix leaking taps and cisterns);
2. improve efficiency of appliances (e.g. AAA efficiency-rated appliances, mulching); and
3. introduction of new methods and technology (e.g. greywater recycling, swales).

Whether these are appropriate to the particular setting and can be adopted requires:

a) an understanding of the current situation;
b) raising awareness at an individual and community level for improved practices;
c) trialing a range of new methods and technologies; and

d) changes to government policies and regional indigenous capacity building.

The matrix suggested below in Table 15.1 is an example of what may be appropriate for one particular community.

15.4 THREE CASE STUDIES

Although urban areas may not be the example or source of innovation for water conservation, there are precedents across Australia that are well worth considering (McLaren *et al.*, 1987). For example, the Western Australian Water & Rivers Commission's water efficiency program in Kalgoorlie (Botica & White, 1996) involved the supply of dual-flush toilets and water efficient fittings to all houses to avoid water use outstripping the supply capacity of the 600 km pipeline from Mundaring Weir. This has become a world-class project with millions of dollars worth of water and energy having been saved after investing only some 2 million dollars. Other examples include: lagoon effluent reuse for sports ovals occurs in many country towns of WA; dual-flush toilet cisterns now required on all new houses in WA; the use of mulch in home gardens, municipal tree planting and roadside revegetation is now commonplace throughout the Perth metropolitan area and many WA regions; the WA State Government has now permitted domestic greywater recycling on a trial basis only for 12 months in three shires. However, there is still an immense potential for change before water use in Australia becomes anywhere near sustainable.

We will now focus on achievements in remote indigenous communities of WA. Using the vehicles of three key case studies: the simple tap (as an example of lifestyle and local maintenance); the evapotranspiration system (as an example of improved efficiency); and the more detailed design input for water harvesting as part of landscaping in the house yard and wider settlement (as an example of a new technology for remote indigenous communities).

Where a standpipe, or any other water supply fitting, has been provided without adequate drainage, ponds form. Dogs and children then use this as their recreation - so does disease. A solution for the standpipe has been a gravel filled sump below the tap to contain leaking water from taps left running. Ceramic disc and anti-vandal taps have also been used. These have often failed because the spring-loaded mechanism wears out due to high use. Such taps can only be fixed with proprietary tools, which the local maintenance worker usually does not have. These taps are very expensive as are replacement parts, which are generally not readily available in regional areas. The standard brass tap is very easy to repair as well as being of low-cost. Highly mineralised water can cause these taps to fail very quickly if not maintained. Initial leaking from the mouth needs only a washer replaced, but if this is not done soon enough the seat below the washer will need to be reground with a hand-held reseating tool. If a tap leaks from the stem of the handle all it needs is a new O-ring to the stem - once again readily available in regional centres. Such tasks can easily be completed at low cost by the householder or the Environmental Health Worker in an indigenous community. A leaking tap can result in the loss of 200 litres/day and a number of taps represent a major waste of water. The standard brass tap, combined with a standpipe sump (or used throughout the dwelling) and an effective local maintenance program, is a low-cost and easy-to-maintain water supply solution.

Table 15.1 Matrix of options for water conservation strategies in remote communities

Process	Lifestyle change	Improve efficiency	New technology
Gain understanding	Regional organisation to conduct a survey of typical domestic practices and appliances in its communities	Gather information on interstate and overseas products. Consult with architects and Aboriginal Housing Directorate as to what household fittings are normally specified and discuss opportunities for introducing new water efficiency fittings and appliances	Install water meters on all taps and water supply pumps to determine total supply, total consumption, individual use, functional use and losses in system
Raise awareness	Distribute booklet and video for use within communities and gather feedback	Present product information to regional organizations and communities	Use existing community-based training and employment programs to install water harvesting swales and basins with landscaping works
Trials	Provide a supply of tap washers and training to all householders in one community and evaluate after 12 months	Install water efficient fittings and appliances in one house and measure consumption before and after	Establish and monitor a greywater reuse system on one house using evapotranspiration
Change policy	Water efficient fittings to be specified in all housing contracts	Homeswest and ATSIC to include funds for yard development in all future housing to enable water and energy efficiency	Establish a regional indigenous water utility to provide technical services to communities in concert with other regional organizations

Septic tanks and leach drains are often the solution for wastewater management in many parts of Australia. However, the transfer of this technology directly to remote indigenous communities has not usually been appropriate. Soils in remote areas are often tight, clayey and of low permeability. During times of high visitor numbers, say during traditional ceremonies, dwellings experience high wastewater production and the standard leach drain will overflow. Moreover, it can become blocked with oils and fats. The tendency has been for engineers to recommend sewers to lagoons as the population grows and money becomes available. This has improved public health, at a great cost, but results in enormous waste of water. The evapotranspiration (ET) system (McGrath et al., 1991) constructed to a similar profile as a leach drain (2000mm wide x 800mm deep) but with graded gravel and sand, instead of backfilling with the local soil, does not rely on infiltration into the surrounding soil. Evaporation at the surface and transpiration from trees and shrubs planted on top and around provides improved performance as well as irrigation for revegetation for shade and shelter or fruit trees. These systems were initially constructed to the same dimensions as leach drains in communities across WA and no overflow has been recorded at any site. In several current research projects the ET system has been redesigned with a shallower profile (600mm deep, 300mm wide at base and 2000mm at top) to improve the success of revegetation efforts. Performance will be monitored. It is expected that sewerage costs can be reduced, plantings made more productive, and enormous water savings made with the use of ET instead of lagoon systems.

Rainwater harvesting in the landscape to improve revegetation efforts and to simplify and reduce the costs of stormwater drainage has been tried in regional areas of WA (Western Australian Water Resources Council, 1986). This technique has been integrated with the close packing of plants using permaculture principles for more productive house yard and settlement landscaping designs (Orion, 1996) and may result in lower water use for the same levels of productivity. The technique has been researched at one site in the arid sub-tropics of WA over the last 12 years, implemented at one indigenous community further north, and has been designed as part of settlement upgrades in several other communities in the region. This more detailed approach to water harvesting requires detailed design on site first followed by earthworks as the first part of the landscaping works. The design requires a series of shallow retention basins cascading their discharge into increasing sized basins and finally discharging off site. The design must integrate ET systems and 'sector analysis' (Mollison, 1988) of solar paths, prevailing winds and views required for the most effective placement of windbreaks, shade trees and food species. Combined with suitable species, considerable water savings can result. Further research is necessary to quantify savings. Of concern is the requirement to keep soil moist in some regions to avoid termite attack on the roots of some exotic species, so resulting in a higher water demand for these planting systems. However, given their success over other revegetation attempts in these difficult regions, it can be argued that the shade, shelter and fresh fruit provided by these systems result in improved health and nutrition and are an effective and justifiable use of limited water supplies. This is particularly the case if the system is made in conjunction with a community-wide water conservation program that results in savings in other areas.

15.5 CONCLUSIONS

Water conservation for the effective management of limited water resources in remote indigenous communities can occur through implementation of a wide range of strategies. Changing lifestyles coupled with improved management and maintenance; the introduction of more efficient technologies; and the use of a holistic design approach to house yards and the settlement as a whole encompass a broad range of techniques and systems that must be considered for both sustainability and improved public health.

15.6 REFERENCES

Botica R & White S (1996), Kalgoorlie-Boulder. *The Water Efficient City*, 23, (5), 1417, Australian Water & Wastewater Association, Sydney.

Flood, J (1983), *Archaeology of the Dreamtime*, Collins, Sydney.

McGrath D.R., Ho, G.E. & Mathew K. (1991), The Effectiveness of Evapotranspiration Systems in Disposing of Wastewater in Remote Aboriginal Communities in Northwest Australia, *Water Science and Technology* Vol. 23, Kyoto, pp 1825-1833, IAWPRC, London.

McLaren N, Heath D. & Ahilefs M. (1987), *Water Conservation for Communities in Arid Areas of South Australia*, Engineering and Water Supply Department, Adelaide.

Mollison, B. (1988), *Permaculture Designer's Manual*, Tagari Publications, Tyalgum, NSW.

Orion, H. (1996), Close Packing of Plants in Water Harvesting Systems, *International Permaculture Conference VI: Designing for a Sustainable Future* Permaculture Association of WA, http://www.cowan.edu.au/-12austin/iRc6/.

Pholeros P., Rainow S. & Torzillo P. (1993), *Housing for Health: Towards a healthy living environment for Aboriginal Australia*, Health Habitat, Sydney.

Race Discrimination Commissioner (1994), *Water - A report on the provision of water and sanitation in remote Aboriginal & Torres Strait Islander communities*, AGPS, Canberra.

Western Australian Water Resources Council (1986), *Water Conservation Through Good Design*, WAWRC, Perth.

16

Groundwater balance in mining

16.1 INTRODUCTION

The groundwater component of the water cycle, which takes place outside the process plant on mine sites can be quite complex. Water management is of primary importance to mining and mineral processing for the successful implementation, operation and ultimate decommissioning of the site. In both arid and humid areas, groundwater can be the largest component of the water cycle. Groundwater is pumped for various purposes, used in processes and for conveyance, and water from different sources may be returned to the ground.

Groundwater for ore processing, tailings transport, for control of dust on haul roads and stockpiles, as well as supplying potable needs, is sourced from pit dewatering or from specially installed bores. Groundwater used may have a range of salinity, depending on the local conditions and the requirements for process and potable water. Water may seep back into the groundwater table through open pits and tailings storages.

16.2 WATER SOURCES

Groundwater is the main source of water for mines in arid Australia. Unlike surface water it is dependable and generally predictable, within the limits of the science of hydrogeology, and the cost of investigation and acquiring the necessary data. Groundwater is generally extracted through specifically designed drilled bores, but also may be pumped from shafts and sumps.

Bore yields range from several hundred kilolitres per day in fractured rock aquifers, to several thousand kilolitres per day in thick sedimentary aquifers. Bore yields may decrease with time as the water table or potentiometric level is drawn down, or fractures become dewatered. Alternatively, yields may increase in in-pit bores as the pit is deepened. Stand-by bores are usually installed to maintain supplies during pump maintenance or breakdown. In-pit dewatering bores must be progressively replaced as the pit is deepened.

Borefields may be completely isolated from the mining operation. They may be on-site but draw from a deep aquifer, which is unconnected to the surface, or they may be located at a considerable distance away from the mine site. However, it is common that borefields are hydraulically connected to the mining operation in some way. Many ore deposits are more permeable than the country rock and therefore are suitable aquifers.

Open pits can be excavated down to the water table with little groundwater impact, however when the operation proceeds below the water table the hydrogeological regime is altered. Dewatering of the pit must be carried out to allow access for vehicles and equipment, to reduce drilling and blasting costs, to reduce water content of ore, and to improve slope stability and mitigate floor heave. The pit may be dewatered from outside the pit with bores or from within the pit by bores or sumps.

Underground workings where water enters from fractures or fissures also require dewatering. Coal seams are usually associated with sandstone aquifers with considerable groundwater content. In karst limestone the presence of groundwater may be a limiting factor in allowing mining to proceed.

To prevent recirculation, excess water from dewatering must be discharged at a significant distance from the pit or underground workings.

Where mineral sands occur below the water table mining may be carried out by dredging, involving an open water filled pit.

Ore transported from the pit to the process plant may also contain a significant component of water. Residue in the form of fine tailings or slimes are piped to residue areas together with a significant water content.

Rainfall events can cause significant runoff to pits, and add to water in residue areas.

16.3 WATER SINKS

Water may be lost from the mine site in a number of ways, evaporation being the most important. Water used for dust control is completely lost to the atmosphere. Water is also evaporated from tailings impoundments, either directly from open water, or from wet material on the beach. Where the water cannot be recirculated into the process, the area of evaporation from tailings needs to be sufficient to balance inflow.

Tailings drain vertically and there may be considerable downwards seepage into the water table. Tailings dams designed to be impermeable may also leak significantly.

Seepage from tailings may recirculate into the aquifer being pumped, although the time scale of most mining operations is short in relation to groundwater flow rates. In arid areas where water conservation is a key issue, the water recovered by underdrainage may offer cost savings over the facility lifetime in excess of the initial construction cost.

Dredge ponds lose water through evaporation, but may also gain water on a seasonal basis from rainfall and groundwater inflow. The seasonal losses will require make up water from borefields.

Recirculation of discharge water may occur through unsuspected pathways from the discharge area to the underground or open cut.

On cessation of mining evaporation from open water-filled voids will take place. This will inevitably lead to an increase in salinity of the water body. The water level in the pit will rise to the point where evaporation will balance groundwater inflow. The void may become a sink in which the salinity continues to rise, and groundwater flow will balance evaporation, or there may be throughflow of groundwater through the pit.

Disturbance to the ground may lead to inter-aquifer flow, through drill holes or disused shafts connecting hydraulically aquifers that were previously separate. This may be done deliberately to assist in dewatering or depressurising.

16.4 WATER QUALITY

We choose Western Australia, a world major mining province, to illustrate our discussion. Groundwater salinity is major factor in Western Australia. Groundwater ranges from fresh in the South West, Pilbara and Kimberley up to hypersaline, up to eight times salinity of seawater, in the Eastern Goldfields.

A mining operation may require separate borefields for different purposes; for instance a small quantity of potable water from a fractured rock aquifer, large amounts of hypersaline groundwater from a more distant paleochannel aquifer. The quality of water used in the process is generally dictated by the local availability. Desalination may be required to supply potable or process water. There may therefore be dual water circuits where water from different sources is used.

The salinity of tailings reclaim water is commonly concentrated from one and a half to three times the salinity of the process water. This water also contains reagents or contaminants.

Groundwater derived from pit dewatering may be acidic due to sulphide oxidation. Groundwater pH is also naturally low in paleochannels in the Eastern Goldfields, and in deep aquifers of the Perth Basin. Acidic waters can be generated from waste dumps, ore stockpiles and residue areas and have the capacity to leach out heavy metals.

The quality of discharge to surface or groundwater is important, as it must meet specified environmental criteria.

16.5 WATER BALANCE

The water balance (see Figure 16.1) consists of quantities, frequency, quality and allowable contaminants of water use for each aspect of the operation. There may be a deficit of volume requiring make up water, or less common, excess water may be discharged from the site. The water balance may change with time as the pit goes deeper and more dewatering is produced.

All the sources and sinks of groundwater are accounted for in a water balance. From the mine planning and management point of view it is important to ensure that dewatering can be disposed of, the process is adequately supplied and that tailings dams have a sufficient area to evaporate excess water.

The overall water balance can be considered in terms of the following components:

16.5.1 Water Gains

- Rainfall
- runoff in pits and tailings dams, dredge ponds, storage dams
- groundwater inflow to pits
- groundwater pumping and surface water from offsite

16.5.2 Water Losses

- evaporation from open water and dust control
- seepage
- discharge offsite

16.5.3 Water Transport

- pipelines and surface flows
- ore
- groundwater recirculation

16.5.4 Recycling

The recycling of water within the mine site has an economic benefit and aims to:

- conserve water
- conserve reagents, and
- minimise volume of excess effluent.

16.6 ENVIRONMENTAL ASPECTS

As well as deriving a water balance for mine water planning purposes, a proper appreciation of the water cycle is needed to assess the effect on the environment.

The likely water balance should be addressed at the approvals stage of a mining operation. The impact of mining operations on the environment and resources (eg water) may be assessable both by mining and by environmental legislation. In Australia proposals deemed to have significant environmental impact are referred to environmental protection authorities for assessment. The level of assessment ranges upwards with increasing public input through a Consultative Environmental Review, a Public Environmental Review, and an Environmental Review and Management Plan, in which commitments may be made on groundwater management.

In order to carry out these commitments it is important to collect the right data. Among the factors that need to be considered are effects of borefields and tailings on phreatophytic vegetation, and the maintenance of groundwater quantity and quality.

Direct discharge to aquifers may be permitted provided there is no adverse impact on environmental value (beneficial use), that is that the pre-existing use of the water is not compromised by any change in water quality due to the mining operation. Discharge to surface water may need to meet specified environmental criteria.

Policy is being developed on voids below water table left after cessation of mining. Some voids can be backfilled during mining operations but most cannot, either because there is not enough backfill or it is not economic. On cessation of mining decant and underdrainage systems for tailings are decommissioned. These should be left in a stable self managing system, sealed and rehabilitated at the surface to minimise leakage of contaminants. Tailings should dry out to a stable moisture content.

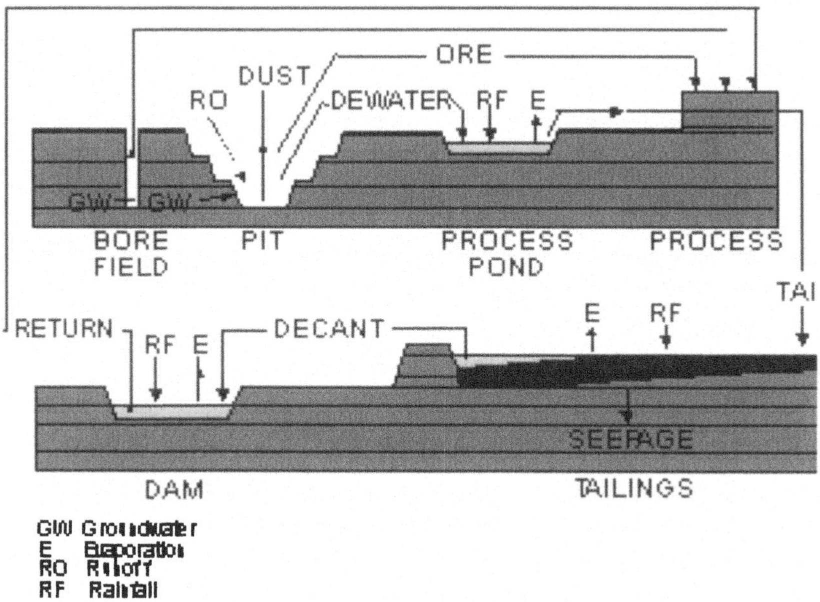

Figure 16.1 Diagrammatic representation of a typical groundwater flows for a mine.

Abstraction and disposal of water is likely to be covered by the relevant water legislation. This may require a monitoring program usually consisting of:

- quantities - monthly pumpage of each bore, metered over certain rates;
- monthly groundwater levels, distinguishing observation or monitoring bores from production bores, noting whether bore is pumping or not, preferably continuous dataloggers; and
- salinity - electrical conductivity measured on water samples monthly from production bores, chemical analysis from pumping bores six monthly.

A water auditor could be requested to develop an actual water flow diagram for a specific mine and to obtain closure for the purposes of confirming compliance with regulations, or for proceeding on the usual path of developing a water management strategy. Auditors with little experience in this field are advised to seek specialist assistance.

16.7 BIBLIOGRAPHY

Netchaef, P. and Dundon, P., 1988. The interaction and impact of water on mining in Australia -an overview: Third International Mine Water Congress, Melbourne, Aust. Inst. Min. Metall.

17

Process: contaminated water systems

17.1 INTRODUCTION

In a variety of arenas, including mining, power generation, mineral processing, refineries, chemical plants, workshops and vehicle maintenance bays, there is a common need for separating from water, liquid hydrocarbons or oil-like substances and sometimes solids. We recognise that minimising the volume of wastewater created (and reviewing the processes creating contamination) is as much a part of a total water management program as is a cost effective treatment system. With stringent regulations governing the discharge of contaminated water now being enforced globally, the treatment of this water and the partial or full reuse of the treated water not only becomes a sound management practice, but addresses broader environmental and water conservation issues as well. This chapter focuses on the more common and readily identifiable contaminated water sources, typically those containing oil, grease and suspended solids. These are a major source of pollution and, if left untreated, impact on the local environment, stormwater or sewer systems.

17.2 SOURCES OF CONTAMINATION

Continuous "point sources" of contaminated water are the easiest to deal with, as they have relatively constant flow with constant content of oil or solids. Point sources include the active use of water used for washing plant and equipment, floor waste generated from maintenance workshops, service and lubrication bays, drainage water from power plant and generator storage areas, and contaminated water from hydrocarbon storage and re-fuelling areas. Water discharging from these areas generally contains relatively high concentrations of oil, and in many instances, varying concentrations of sand, silt and suspended solids. These waste streams therefore require treatment.

The other source to consider is stormwater. In order to minimise potentially contaminated stormwater discharging to the environment, measures must be taken for control and treatment. The concentration of oil in stormwater run off around vehicle parking and lay-down areas, haulage winder houses, crushing stations, and unroofed contaminated areas can reasonably be assumed greater than discharges allowed and are thus a very real threat to drainage systems. It is therefore necessary to ensure contaminated areas are properly bunded or contained, not only to trap the contaminated stormwater within, but to prevent the ingress of uncontaminated stormwater, which does not require treating. Site layout designs should endeavour to keep contaminating functions undercover to avoid having to treat stormwater at all.

17.3 STRATEGIES FOR TREATING WASTEWATER

To determine the most suitable strategy for treating oil, grease and/or water laden with solids, a site water audit or water balance needs to be undertaken. This establishes the specific requirements of the process, allowing a wastewater treatment facility to be accurately selected and designed. The water balance should include:

- the amount of wastewater generated. In the wash-down of haulage trucks for example, this would be the flow rate of the 'water monitor' or pressure cleaner used to wash the equipment.
- the frequency and duration of the wash-down or waste-water generating process.
- the type and area of surface to be drained.
- the storage capacity or retention volume of holding sumps.
- the peak flow rate from a storm. Consideration must be given to the management of rainfall that comes in contact with an oily contaminated area subject to rainfall entry, such as an un-roofed vehicle washbay.

- the type of contaminants. Eg oil, solids, degreasers/detergents/ emulsifiers.

Once a water balance has been undertaken, a number of questions then arise.

17.3.1 Is it necessary to treat all of the stormwater that falls on a contaminated area?

Contaminated stormwater can be divided into two categories:

- stormwater polluted from contact with a 'high risk' contaminated area such as an unroofed wash-bay.
- stormwater polluted from a low risk contaminated area such as a vehicle parking area.

From an environmental point of view, the safest course of action is to treat all of the stormwater. Whilst this may be practical with the use of retention interceptors (in- ground tanks or pits sized to capture all the rainfall run-off from the contaminated area) in low risk, hard standing areas (eg. carparks), large and costly treatment systems are needed to process stormwater draining from all but the smallest high risk areas. Alternatively, large holding ponds could be used to accumulate peak flows for processing at lower flow rates over a longer period of time.

17.3.2 What criteria should be used to determine how much water to treat?

A treatment system can be designed for a specific flow rate or a specific storm event, with any excess simply accumulating on the contaminated area. Whilst this may be practical for bunded areas that have controlled release of polluted water (eg a diesel storage facility), it is generally not practical for areas of high usage (eg wash-bays) or unroofed areas subject to high rainfall.

An alternative method is to capture and treat the 'first flush' of contaminated stormwater from the contaminated area, allowing clean uncontaminated stormwater to bypass the captured contaminated water. The 'first flush' hypothesis assumes that contaminants present on the area will be washed off the area in the early stages of a storm and further run-off will not impact on the environment or receiving waters. The quantity of rainfall required to do this is generally accepted as the first 10mm or 15mm falling on the contaminated area. Further rainfall after this should be relatively free of contaminants and therefore ought not require treatment. By containing and treating the first flush and allowing the clean uncontaminated storm water to bypass it, a practical low flow and cost effective treatment system can be installed.

In the design and sizing of a first flush stormwater bypass system, a number of factors need to be considered including:

- type and porosity of surface. An old bitumen surface will release oily contaminants more slowly than will a new concrete surface.
- fail safe bypass designs. Bypassing non turbulent flows will reduce mixing of oily water with clean water.
- location of drains; whether the area drains centrally or to a point on the perimeter. Contaminated run off from an area that drains to the perimeter will have some water that takes longer to reach the drain than contaminated run off from a centrally draining area.
- potential environmental risk. If there is risk of contaminated water bypassing collection, then the ultimate destination of the bypass water must be considered. An on-site evaporation pond may be more suited to receiving bypass water than allowing direct discharge to ground seepage or the environment.
- specific requirements of the regulatory authorities. Some regulatory authorities may only allow a bypass system providing there is no practicable or possible alternative. The relevant regulatory authorities should be consulted to determine the acceptability of any stormwater bypass system.
- how to bypass the clean water. There are a number of storm bypass 'systems' available, ranging in complexity from simple pipe tee's, special pits with automatic valving, through to concrete and glass reinforced plastic separation interceptors, with built in bypass weirs.

17.3.3 How should the water to be treated be segregated from the water that is not to be treated?

While current site practices will be hard to change or modify, it is essential that good housekeeping procedures be adopted, not only to simplify but to reduce the cost of a treatment facility. Drainage from areas such as roofs, lawns or other areas not normally expected to contain hydrocarbons should be diverted away from any treatment facility. This can be achieved by minimising contaminated water entry into a treatment plant by incorporating a physical barrier or 'bunding'. This will not only contain the contaminated water within, but to prevent ingress of additional clean stormwater from other areas. The bunding may be in the form of a wall, a concrete kerb, a ramp or hump if vehicle access is required, or even the natural contours of the land.

17.3.4 How can the size and cost of a treatment plant be minimised?

By understanding how a treatment plant works and the conditions that cause overloading or problems, good 'housekeeping' procedures can be adopted to reduce the cost and size of the facility. These include:

- utilizing waste oil storage facilities to dispose of used oil (eg. from gearbox, engine and filter changeover) rather than tipping into the treatment plant drainage system where it will be re-separated (unnecessarily) from the water.
- Using oil recovery equipment such as belt or floating skimmers will help remove large 'slugs' of free floating oil prior to the wastewater entering a treatment plant. Slugs of free oil rise rapidly to the surface where they need to be recovered, rather than left in the system to be processed through a treatment plant. This helps reduce overloading of the treatment plant.
- storing of oil drums on containment pallets, preferably under cover.
- directing multiple source contamination to a centralised collection and treatment point (where drainage and distance permit).
- moving oil and fuel storage/dispensing facilities to a central point and chemical storage to another.
- reducing and eliminating point source contamination with failsafe containment and entrapment, such as using oil containment booms on waterways, ponds and sumps and using oil absorbent booms and materials around specific work areas.
- choosing suitable types of chemicals. Chemicals such as degreasers and detergents combined with oil and water creating emulsions. Strong or stable emulsions may pass through gravity based treatment systems, without the oil being removed. This necessitates a more complex and costly treatment facility. The use of minimal amounts of 'quick breaking' degreasers and detergents allows the emulsion to destabilise (the bond between the oil and chemical weakens), releasing free oil droplets. These free oil droplets can then be removed in a gravity-based treatment system.

17.4 SYSTEMS AVAILABLE FOR SEPARATING OR RECOVERING OIL FROM WATER

A number of systems for separating and recovering oil from water are available, ranging from simple holding ponds, enhanced gravity separators, skimming devices, hydrocyclones, media filters, chemicals and membrane technology based systems.

Skimming or oil recovery equipment is used to recover free-floating surface oil, not to separate oil from water. There is very little value in installing an oil-skimming device when the wastewater contains small micron oil droplets dispersed throughout it. In this instance, the oil droplets have not sufficiently coalesced into larger droplets that are able to rise to the surface, form a 'slick' or 'slug' and be recovered or skimmed. A separation system rather than a skimming system is therefore required.

Traditionally, the simplest and least costly 'separators' have been used, although by today's standards, they have generally been inadequate in removing oil so that water is suitable for discharge to sewer, ground seepage or for re-use. Growing environmental awareness, combined with mounting pressure from regulating authorities and a commitment from industry to comply with environmental guidelines, has helped to replace traditional outdated 'triple interceptor' style pits (see Figure 17.1) with later technologies, that achieve higher oil removal rates.

Achieving higher quality treated water, increases the cost of the treatment facility. Costs include capital equipment, consumable chemicals, skilled operators or energy inputs. We will focus on methods that provide adequate separation, with minimal cost and maintenance, so that the treated waste water can be used for dust suppression, wash-down or other processes using a water stream of low order quality.

17.4.1 Baffle Separators

The simplest possible separator is an empty chamber sized to contain spills. Such a spill control separator is too small to intercept small, dispersed oil droplets and is really only suitable for intercepting large oil spills. Even then, unless the accumulated oil is removed regularly, it becomes less effective, particularly during high rainfall periods when accumulated oil can be washed through the system. Turbulence in the water and the distance an oil droplet has to rise to reach the surface limit baffle separators in their separation ability. The Triple Interceptor Pit or Petrol and Oil Trap (P&O Trap) are both variations of a spill control separator and have, for many years, been the main stay of hydrocarbon interception (see Figure 17.1).

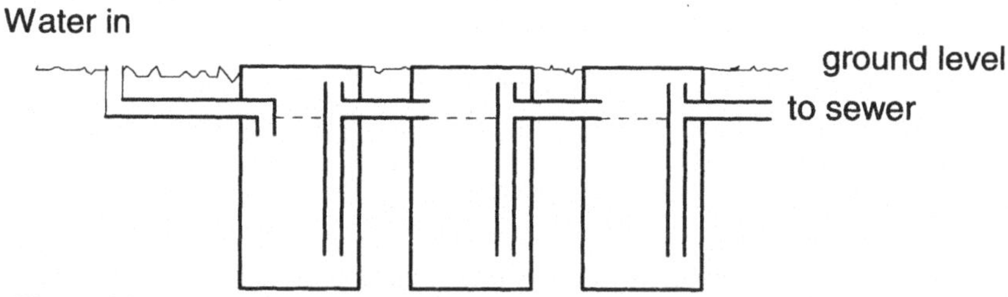

Figure 17.1 Triple Interceptor Pit. Oily water flows in with the oil progressively separating in each chamber by gravity alone. The cleaner water flows under the oilier upper water to the next chamber.

API (American Petroleum Institute) separators are gravity separators, similar to spill control separators, but generally larger, more effective and often equipped with oil removal facilities. These separators are used extensively in oil refineries and chemical plants where water streams containing large amounts of oil are present (see Figure 17.2).

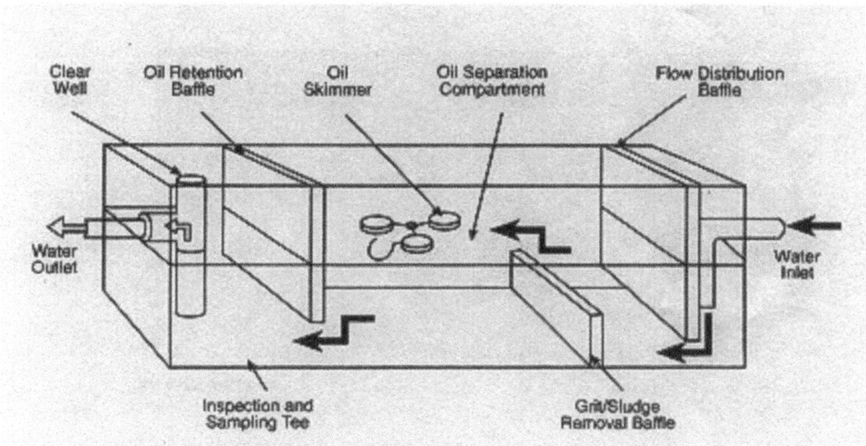

Figure 17.2 API (American Petroleum Institute) separator.

The advantages of baffle separators are simplicity of design, low cost, low maintenance and resistance to plugging with solids. Oil separation is achieved solely by gravity in near calm water. This results in large tanks with low flow rates and the quality of effluent a function of residence time and/or tank volume. Their disadvantage is the relatively poor quality of separation compared with later technologies and the inability to achieve effluent qualities that comply with current regulations.

17.4.2 Enhanced Gravity Separators

Various types of enhanced gravity separators have been used to provide better separation quality than API or Baffle gravity separators, while still maintaining low capital and maintenance costs. These separators 'enhance gravity' or speed up the process of oil and water separation with a closely spaced inert medium or plate that increases residence time compared with that in either 'pure gravity' or API separators. Designs that have been successful include inclined or tilted plate, horizontal corrugated plate and multiple angle plate separators. This group of separators is commonly referred to as Coalescing Plate Separators (CPS) or Coalescing Plate Interceptors (CPI).

Operation of a CPS is simple. Oily wastewater is introduced into an inlet chamber (see Figure 17.3) by either gravity flow or a pump. An inlet weir separates this chamber from the plate pack section. Heavy solids settle out and 100% oil slugs rise to the surface. The remaining oily water mixture flows through a stack or stacks of closely spaced plates, where finer solids progressively settle.

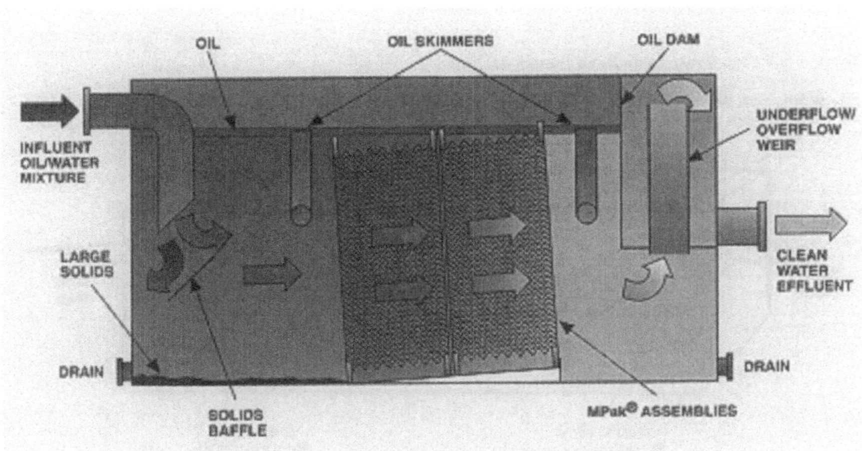

Figure 17.3 Typical Coalescing Plate Separator, courtesy Facet International.

Small oil droplets are progressively separated through the plates, where they coalesce with other droplets and rise to the surface. An oil dam prevents the separated oil from entering the outlet weir and oil skimmers allow the separated oil to be skimmed to a collection vessel The treated water underflows the oil dam and over an outlet weir to discharge (see Figure 17.3).

Figure 17.4 Multiple Angle Coalescing Plates, photo courtesy Facet International.

Multiple Angle plate separators (see Figure 17.4) were developed to take advantage of the virtues of older style plate separator while eliminating potential plugging disadvantages and the inconsistent plate spacings. With plate corrugations in both directions the surface area to volume ratio is extremely high. This combined with plate spacings of either 6mm or 12mm, enables small oil droplets down to 20 microns in size to be removed in very short periods of time. The plates are designed to shed solids quickly to the hopper of the separator, thus avoiding plugging.

Figure 17.5 Above ground Coalescing Plate Separator, heavy vehicle washpad, Western Australia.

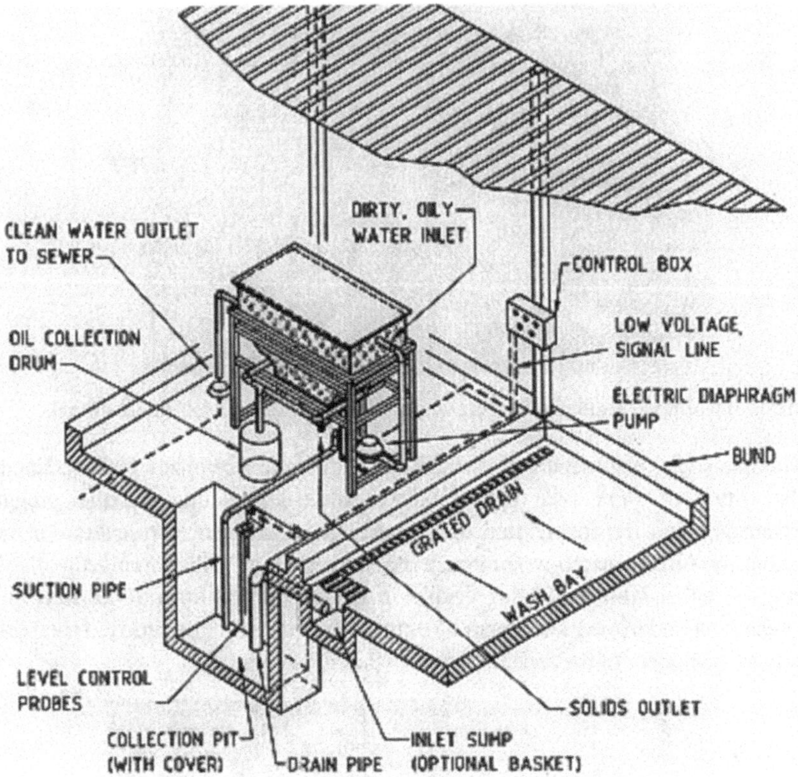

Figure 17.6 Typical Coalescing Plate Separator system, general arrangement drawing.

Such plates can be installed in either above ground tanks with the wastewater pumped into the tank (see Figures 17.5 and 17.6) or in below-ground concrete pits. They may also be retrofitted to existing systems such as API or baffle pits to substantially improve the effluent quality of the discharge water.

17.4.3 Dissolved Air Flotation

Dissolved Air Flotation (DAF) is a physical/chemical process in which microscopic air bubbles become attached to solids particles suspended in a liquid, causing the solid particles to float. This flotation method is well suited for clarification of industrial wastes where the specific gravity of the waste contaminants varies. When preceded by an effective waste-conditioning (chemical treatment) step, DAF will effectively remove free oil, light and heavy solids and emulsified oils. Waste conditioning chemicals such as coagulants, flocculants, acids and bases are typically used in the removal of these contaminants.

The chemicals demulsify the oil and DAF floats the flocculated material to the surface where it is skimmed off in the form of sludge. While a DAF system has a high capital cost, can be maintenance intensive and requires in-house expertise, a variety of chemical inputs and sludge disposal, it is effective where large, continuous volumes of wastewater require treatment. In particular wastewater, containing floating, sinking and suspended solids as well as oily water that becomes emulsified from the use of degreasers, detergents or surfactants, can be clarified to high levels. Wide variations in influent conditions (eg oil slugs) should be avoided.

17.4.4 Chemical Treatment

Chemical treatment lends itself to applications with complex, hard to treat, low volume wastewaters, typically those containing emulsions of oil, colloidal particles, heavy metals, phosphates, suspended solids and water. While the chemical treatment regime may be similar to that of DAF, the wastewater is usually processed as a batch, with the resulting contaminants settled as a sludge. This type of treatment system requires multiple dosing of different chemicals (including pre and/or post pH correction), as well as in house expertise to operate it effectively, regular equipment calibration and continual purchase of consumable chemicals.

17.4.5 Membrane Technology

One of the more recent developments in handling oil in water emulsions is the use of Hydrophilic Membranes. Wastewater is pumped through a hollow fibre micro-filtration membrane, which encourages the formation of a static water layer against the membrane wall. This effectively concentrates the oil emulsion so that the stability declines and free oil particles begin to form leaving a highly concentrated sludge and a clean permeate. Membrane systems are suitable for treating low volume oil/water/surfactant emulsions where free oil/solids have already been removed. The treated water is usually of sufficient quality to recycle or to discharge to sensitive ecosystems. Membrane systems do tend to be expensive to purchase and to maintain and require regular cleaning to prevent membrane blinding or fouling.

17.5 SELECTION AND DESIGN OF COALESCING PLATE SEPARATORS

Stokes Law enables us to predict the rate of rise and fall of a droplet or particle in a laminar fluid stream and this determines the performance of coalescing plate separators. Separator systems that are properly sized and designed guarantee that the discharge quality will conform to regulations. Proper design is imperative when the permissible oil content in discharge water is below 10 parts per million. Among factors which must be considered in the design and selection of an oil separation system are:

- flow rate (gravity or pumped, type of pump).
- degree of separation required (effluent quality).
- amount of oil in the water, specific gravity and temperature.
- existing pipework, valves, fittings etc.
- degree of emulsification, either mechanical or chemical, of the oil.
- oil droplet size and distribution.
- removal method of the recovered oil.
- vessel (tank or pit) designed correctly for laminar flow.
- solids loading.

All the above parameters are necessary to accurately design a coalescing plate separator system, yet the key to system sizing is the actual size of oil droplet to be removed. This in turn is influenced by the degree of shear (or mechanical emulsification) of the oil droplets. The most cost effective coalescing plate separator is one that will remove an oil droplet down to 20 to 30 micron in size.

Most physical mixtures of oil and water will separate given sufficient time. The smaller the oil droplet or the closer to water the droplet's specific gravity is, the longer the separation can take. From Table 17.1 it can be seen that a 60 micron droplet will rise approximately nine times faster than a 20 micron droplet. Equipment or conditions that produce small oil droplet sizes in the inflow will therefore increase separator size to allow for the additional time required for the smaller droplet to be removed.

Conditions causing small droplets are those that produce shear in the incoming flow, with the following all reducing droplet size, more or less in order of severity:

- pumps (especially centrifugal), valves (especially globe), pipe restrictions and unduly small influent pipes. Low shear, positive displacement, pumps at low speed must be used when pumping to a separator;
- emulsifying agents such as detergents, degreasers and other surfactants contribute to smaller droplet size. A chemical emulsion consists of droplets of one immiscible fluid dispersed in another continuous fluid. Coalescing Plate Separators enhance gravity to separate free oil from water and are unable to separate chemical emulsions brought about by the usage of detergents, degreasers and other types of chemicals.

Table 17.1. Droplet 'Rise/Fall' time. Table courtesy Facet Quantek.

Droplet Rise/Fall Time. Travel Time For 75mm Distance @ 20°C (Hr:Min:Sec)

DROPLET DIAMETER (MICRONS)	SETTLING TIME FOR DIRT, SG 2.0	OIL RISE TIME S.G. 0.85	OIL RISE TIME S.G. 0.90
300	0:00:02	0:00:12	0:00:15
150	0:00:06	0:00:42	0:01:03
125	0:00:09	0:01:00	0:01:27
90	0:00:17	0:01:54	0:02:54
60	0:00:39	0:04:12	0:06:36
50	0:00:56	0:06:18	0:09:18
40	0:01:27	0:09:36	0:14:24
30	0:02:35	0:17:22	0:25:48
20	0:05:49	0:38:46	0:58:08
15	0:10:20	1:08:54	1:43:22
10	0:23:15	2:35:02	3:52:33
5	1:33:01	10:02:09	15:30:14
1	38:45:35	258:23:53	387:35:49

17.5.1 Settling of solids

Oil droplets coalesce into larger spherical droplets, while solids tend to agglomerate into larger masses rather than coalesce into particles that have lower surface/volume ratios like oil. In applications involving the washdown of trucks and earthmoving machinery, the reduction of solids from the water stream is as important as the removal of hydrocarbons. A ramped pit or sump, with sufficient residence time to allow solids to settle and access to a site vehicle to clean it out is commonly referred to as a 'Drive-in sump'. This acts as a solids pre-settlement chamber prior to a treatment plant, and should aim at removing the bulk of the solids and particulate matter in an area that is readily accessible for easy removal. This in turn allows a treatment plant to work efficiently to remove oil and finer suspended solids without being overloaded. The design of silt removal systems is site specific and dependent on the amount and nature of the solids.

17.6 OPTIONS FOR TREATED WATER

The final destination for treated wastewater needs to be addressed within a water management program. All too often the treated water is merely seen as a 'waste stream' only suitable for discharge off site or destined for evaporation ponds. Where quality available water is costly, or off site disposal costs are high (or not acceptable), this resource can be utilised for low level re-use in areas such as dust suppression and plant and equipment wash-down. With the addition of polishing filters (such as membranes and absorption or adsorption media filters), this low level re-use may be extended to other areas that require a higher quality of treated water.

17.7 SUMMARY AND CONCLUSIONS

Oil and grease in stormwater continue to be a problem globally, with environmental bodies around the world setting standards on maximum allowable contamination of effluent water. These standards, along with site practices and procedures, influence the types of treatment systems that are suitable or acceptable.

The key to managing a Contaminated Water System is in the segregation of oil contaminated streams from clean stormwater and other waste streams, the minimisation of point source contamination and other strategic site housekeeping procedures. These measures help to lower the influent oil concentrations and solids loading to a point where the contaminated water volume is economically viable to treat.

The broader community increasingly expects industry to return its wastewater to the environment in the same quality as it was received. Government resolve, public awareness, costs to industry and treatment technologies available are all factors to consider in managing a contaminated water system.

17.8 BIBLIOGRAPHY

Baldwin Industrial Systems, Sydney, "*Private correspondence with Mr R Baldwin and Mr D Fletcher concerning Coalescing Plate Separators*", 1996-1999.

Conder Products, *Interceptors-The Complete Guide*, Bulletin 52Xn8, May 1997.

Facet International, *Engineering Clear Solutions for Environmental Clean-up*, Facet International, Inc., 1994.

Facet Quantek, *Coalescing Plate Separators for Environmental Clean–up and Resource Recovery*, Publication 3-91/5M.

Miley, A., "Treatment of oil contaminated water from mining sites", a paper presented at the *Epa/Market Australia Workshop, Mining and the Environment*, Jakarta, Indonesia, June, 1996

Mohr, Kirby S., "A new type of high efficiency oil-water separator for better water quality management", a paper presented at the *Pacific pollution control association annual meeting*, Boise, Idaho, 1992.

Mohr, Kirby S., *Treating stormwater run off for removal of contaminants*, Facet International, Inc. Tulsa, Okalahoma.

NWW AcumenPacific, Optimum MF Microfiltration, *Application Bulletin*, Nov. 1993.

18

Arena: Material transport

Raw material is transported by several means including trucks or road trains, aircraft, rail, ships, conveyor belt or slurry pipeline. The major user of water in this group is the slurry pipeline. Associated with this and the other means of transport is the suppression of dust using water on conveyors or material stockpiles. From a water auditors point of view, therefore, in this chapter we examine both water use in slurry pipelines for mass transfer and the use of water for dust suppression and control.

18.1 WATER USE IN MASS TRANSFER APPLICATIONS

18.1.1 Introduction

Of the various ways water might be involved in the transport of materials we focus here on the use of closed conduits or pipes. As in other arenas of auditing, the preparation of a diagram representing all water movement of the total process is essential. It is important to remember that flow can be continuous or intermittent, even perennial or non-perennial. The discipline of Rheology is the academic discipline that investigates the flow of matter and its deformation and it is important in describing material transport by water. In general particle size, angularity and grading (distribution of particle size) are the important parameters. The raw material is collected, sometimes washed and then mixed with water to a lean, medium or

 Water Auditing and Water Conservation edited by Jeffrey Sturman, Goen Ho and Kuruvilla Mathew. ISBN: 1 900222 52 3. Published by IWA Publishing, London, UK.

dense phase. The dense phase has been proven workable and has the associated potential for high energy and water usage efficiency. Pumps are now available to transport dense phase slurries. Of the various types of pump available, piston and diaphragm pumps are superior in terms of energy usage, capital cost and life expectancy. Two systems are possible: a single delivery pipeline or a recirculating system. Recirculation of water becomes redundant with the dense phase.

Mass transport of any solid material by water is one of the most common industrial processes today. The water is not just slurried with the material; it can be used in any proportion of say less than 0.1% to as high as 85% w/w. The water can be in either gas or liquid phase, as steam is often used to transport material, especially in the new process technologies that use pressure autoclaves for extracting metal cations from dilute minerals.

To audit a system you need to understand the process, and the physical and chemical conditions in the circuit. Assessing the efficiency of the process and its use of precious resources is a very important function. In Figure 18.1 we show a typical material and water flow diagram for material transport. British Elecricity International (1990-1992) and Jacobs (c.1991) give further information about power station practice in particular. Most Chemical Engineering and Civil Engineering departments are involved in some form of materials transport research. In this chapter we will deal with the practical applications of material transport using water in its liquid phase, we will consider the following:

- the chemical and physical processes that take part during the process;
- the materials that are used;
- mass and energy balances (this is a rigorous and often complex study); and
- reclamation of the supernatant water.

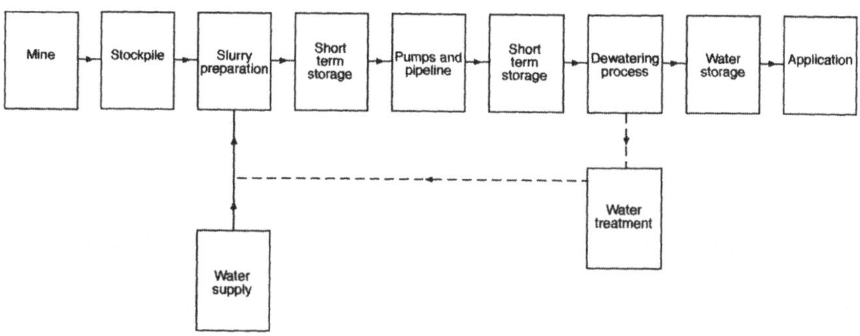

Figure 18.1 A typical material and water flow diagram for transport of a slurried mineral from a mine.

The auditor in my opinion should always look at recovery of the water, even if it may appear to be sub economic, as it frequently becomes an issue of great importance during the life of a mine or an industrial process. The water auditor can anticipate gains by upgrading the transport medium in the direction from lean to medium to dense phase, often with modest capital outlays. Water losses also occur from entrainment in the material and by evaporation. Finally dust suppression systems might provide opportunities for the auditor to consider so as to save water. (This topic is treated below.)

Coal Fired Power Stations in many parts of the world have major expenditure invested in slurry of waste material, in particular, fly ash from the central plant to be piped to some location where the ash is stored. For each 100,000,000 tonnes of coal approximately 20,000,000 tonnes of ash needs to be disposed of in ash slurry given an average ash content of over 18%.

18.1.2 Physical and Chemical Considerations

18.1.2.1 Chemical Considerations

Let us examine the reasons why water is used as the medium and process fluid for material transport. Water is an unusual liquid. If we glance at the periodic table we find that the elements other than oxygen in-group 6a are Sulphur, Selenium and Tellurium. The corresponding compounds to water, i.e. Hydrogen Sulphide, Selenide and Telluride are all foul smelling highly toxic gases. Something about this compound Water is unique and valuable. Other compounds such as HF, HCl, HBr and HI are all elements of the same group or family and are similar chemically. This is not the case with water. Part of the answer can be found in the fact that hydrogen bonding does make the properties of water unique. Some other factors are:

- its liquid phase at ambient temperature;
- its high boiling point;
- high latent heat of vaporisation and fusion;
- universal solvent properties;
- chemical inertness; and
- solid is less dense than the liquid

Water interacts with the material it is transporting. It hydrates some materials, such as fly ash, which is totally anhydrous when first produced and thus changes the properties markedly. Also, many minerals including those from fly ash from combustion of coal dissolve in the water changing the chemistry from neutral to either highly acidic or highly alkaline. This aspect has a very large effect on the water auditor, as disposing of the resulting water can be difficult and highly polluting.

Water from various sources can never be expected to have the same properties, essentially because of what is dissolved in the water. Water can be thought of as being in many energy states. The highest Energy State is pure demineralised water. To get to this state you must put energy into the water to remove the dissolved material. To move materials like paint emulsions, pharmaceuticals, and highly pure materials such as pigments and ion exchange resins, you must use water of this quality. An audit of the water balance in such a circuit becomes an accounting exercise as water of this quality can cost up to US$10 per kilolitre. This may seem trivial, especially when you buy demineralised water from the supermarket for US$200 to US$500 per kilolitre, but this is an average price to pay for such high quality water. Water that is of 200 ppm TDS or less, if you can obtain it from a water service provider will cost up to several US$ per kilolitre. Water used for transporting material in the Western Australian Goldfields can have a salinity of approximately 40,000 to 100,000 ppm; that is up to three times as saline as sea water.

18.1.2.2 Physical Properties

The physical properties of water that concern us are:

- the temperature limitations that can be imposed on the fluid;
- the electrostatic and physical effects such as Zeta Potential; and
- the quality of the water.

The electrostatic properties can be attributed to the concentrations of certain salts. The effects are complex and often difficult to interpret, having a range of effects on material transport. The Zeta potential is simply the charge on a particle that is suspended in the fluid referenced to the surrounding particles. If the charge is similar you can expect the material to stay in suspension and behave more like a homogenous fluid than a two-phase mixture. If the charges are dissimilar then the material will precipitate. Such materials as water based paints flow like this. It is most important to have particle size graded into very discrete sizes (within sharp tolerances) so that the material flows smoothly and does not separate. To avoid separation the correct quality water is used with added chemicals called dispersing agents. For example Sodium acrylate with a molecular weight in a certain range will ensure that the particles in paint do not precipitate.

Attempts have been made to use magnetic fields applied to the water slurries to somehow change the properties of the water to prevent corrosion and to stop precipitation. Every few years some company will come up with a very expensive device that consists of a magnet suspended over the pipe. This device has been sold to prevent all manner of problems with water borne transport. Scant evidence exists that support any favourable results; yet many gullible operators have applied the devices, as they appear to provide a cheap panacea to a range of complex technical problems.

18.1.2.3 Water Quality

Ion exchange resins are polymers that are made to uniform size. They are manufactured with sulphonate or ammonium functional groups attached to the polymer. The resins are either positively charged in the case of Anion Resins or negatively charged as in the case of Cation Resins. The resins are moved in slurries and pumped around circuits. The differences in specific gravity make separation easy and regeneration possible. All resins are slurried using the highest quality water to prevent contamination. Ion exchange resins are designed to be slurried using high quality water. The separation is most important as these materials cost up to US$3000 per kilolitre.

18.1.3 Properties of the Solid Materials

Slurries can transport many materials; some of the most elaborate systems involve coal slurries, transported in pipelines of up to hundreds of kilometres in length. We choose coal as a representative material in slurry transport. The transport of coal slurry is quite a complex technical exercise and water reclamation is an important part of the whole enterprise.

The properties of the coal are important. The size of the coal and the range of size of the particles make a difference to the way the particles flow in the fluid. For example coal might be crushed to an upper limit size of only 2.5mm before being introducing to the line. The energy required for the slurry to flow down the line is also extensive, for not only does the material have to be lifted over hills, it must overcome friction and drag. The study of material flow and deformation is called Rheology. The important properties of solid materials if they are to be transported efficiently are:

- particle size;
- particle shape; and
- size distribution.

Another property to be considered, though of less importance, is specific gravity.

To achieve the optimum design performance it is necessary to prepare the material for transporting the long distances without material separating out of the fluid and blocking the pipe work. The important condition to achieve is approximately equal material and fluid velocities. This means that drag factors, shape and size, are important.

18.1.4 Coal Transport and Washing

The water that is used to transport the coal slurry must have some utility at the end of the process. Water auditing and conservation measures are essential. The coal is transported from the mine, after processing, to the user and in most cases this is a Power Station. Coal is still the major fuel in the production of power and coal is often cheap, reliable and the technology is well understood.

The slurry of coal transported in pipeline is approximately 50% dry weight coal. The remainder is water. In some lines it is necessary to alter the chemistry to avoid corrosion. This may mean the addition of reducing agents to remove the oxygen from the water and hence prevent corrosion of the steel pipe work. Where this is practiced it is a very expensive exercise. Water contains approximately 8 parts per million (ppm) of oxygen under standard conditions, and to reduce that requires the addition of 30 to 40 ppm of a chemical such as Ammonium bisulphite. This chemical is quickly oxidised to Ammonium sulphate and at the same time reduces the oxygen level in the water thus rendering it less corrosive. The chemical addition makes the process more difficult as the water needs to be disposed of effectively. New pipe materials have made this unnecessary now, but the water auditor will be exposed to older pipelines as well as to new.

Coal mined in Australia, for example, can have an ash content as high as 36%. This can create enormous problems of disposal and waste as the inorganic fraction must be heated in the combustion process to 1300°C then cooled down. This is a great waste of energy and often the material fuses. To overcome these difficulties, before the coal is pumped into the line, it is often washed. This means that the organic part of the coal is separated from the inorganic components. The simplest way of washing coal is to make use of the difference in specific gravity of the organic components and the mineral components. Organic material, or the coal substance, has a specific gravity between 0.9 and 1.2. The specific gravity of most rock and minerals is above 2.0. Crushing the coal to a size usually between 50 to 100mm upper size and then floating it in a fluid with an intermediate specific gravity will separate the two fractions. The fluid used for this purpose is water mixed with fine magnetite. The resulting slurry has a specific gravity of approximately 1.6, and this will float the organic components and the mineral components will sink. The magnetite, being magnetic, can be removed from the fluid matrix and be used successively, with only small amounts remaining in the coal. The water is reused with some loss to the coal. The chemistry of the water changes slightly and some make up is needed. The coal can have the inorganic mineral components reduced by up to 70%. The water auditor (or environmental auditor) often focuses on the professional operation of such plants to minimise unwanted discharges.

18.1.5 Slurry Densities

There are three main slurry types:

- lean;
- medium; and
- dense phase.

The density of the slurry is important. Increasing the slurry density, that is reducing the liquid portion, means less pumping energy is required as there is less mass to shift and less water to be found for the process. The corollary is that less water needs to be treated at the collection point and more transported material can be moved down the pipe per unit mass of slurry.

18.1.5.1 Lean Phase Slurries

The lean phase can be as low as 0.5 to 5.0% by weight, with velocities in the pipeline as high as 4.5m/sec. In an industrial process that is not closely monitored, slurry densities tend to be low as control problems are minimised with more dilute slurries. They have a lower effective viscosity and there is less wear on the pipeline. Radioactive Crompton scatter devices measure the radiative backscatter from inorganic material in the slurry, for organic material is often transparent to radioactivity. Such measuring devices, in association with automatic controls, allow a slurry of stable consistency to be maintained. Sewerage is an example of lean slurry, with the potential for reclamation and recycling of the water.

18.1.5.2 Medium Phase Slurries

The most common type of slurry density falls into the range 5 to 50% by weight. The velocities are lower in the pipeline, typically about 2.5 m/sec. The pipe materials for this type of slurry can be normal mild steel or cement lined pipe, fibreglass reinforced pipe, and in some cases high-density polyethylene. Reinforced cement pipe is popular. Most fly ash and slurries from mining fall into this category. Typically the specific gravity of the mixture is 1.3 to 1.6.

18.1.5.3 Dense Phase Slurry

Dense phase slurries are different to the above in that they are very much more concentrated in the solid phase material. The range of solids can be as high as 90% and the material can look like a solid, whereas it behaves as a plastic. The advantages of dense phase slurries are:

- the amount of water required is minimal;
- the material will stack and does not need a dam or any containment area,
- velocities of transport are lower typically 1.5 to 2.0 m/sec;
- pipes are smaller in diameter;
- water collection and treatment is minimal and therefore the capital and operating costs are lower.

These are admirable advantages, yet there are practical difficulties. The material must be thoroughly mixed to get a homogeneous mixture that has uniform rheology. To make the material flow positive displacement pumps are required. These pumps are expensive and require considerable maintenance.

18.1.6 Pump Types

Weir, Warman and the Norwegian Company GEHO, are the three common manufacturers of slurry pumps. The technology is highly specialised. The centrifugal slurry pump is typically one that is robust with special materials for the impeller, usually rubber or ceramic lining, with large clearances between the casing and the impeller. In centrifugal pumps it should be noted that denser slurries can be pumped with the same pump by reducing pump speed and replacing the impeller, thereby allowing higher density slurries and reduced water demand, possibly with energy saving dependent on the rheology. The pump must have a very effective mechanical seal or a special water balance on the shaft gland. Typically centrifugal pumps are used with lean and medium phase slurries, with diaphragm and piston pumps taking over with denser medium phase and dense phase slurries. Diaphragm pumps can be used in conjunction with piston pumps.

Abrasive material can come through the gland and wear a track down the shaft. This can be avoided with good design. Material selection is important. Inert material, some at the cutting edge of technology is necessary in many of the current applications. An example of the limits of slurry pumping capability is seen at a nickel refinery in Australia. In this plant there is a mixture of diaphragm slurry pumps. The pumps have to move a slurry at 190°C consisting of 30% sulphuric acid at a pressure of 5 megapascals and 40% solids by weight. This of course is an application for diaphragm pumps with only the most sophisticated materials coming into contact with the fluid. Materials like pure Tantalum and to some extent Titanium can stand these conditions. Most of the problems with pumps are due to erosion and wear, and the life you can expect from a pump is limited.

18.1.7 Water Collection and Treatment

One of the most interesting problems for a water auditor is what to do with the water after you have separated out the solid material. This water is sometimes called the supernatant or tailings water. In the above example of coal slurry, the end user was a Power Company. A large power station that burns coal consumes prodigious amounts of water in the production of demineralised water, for the cooling tower makeup and in slurry for the fly ash disposal. Auditing and producing an accurate water balance is essential. The water is often changed chemically by intimate mixing with the mineral. With coal you can expect some change in the pH. Most minerals contain pyrites. This common material oxidises readily and releases protons in the process.

$$Fe^{2+} + HOH + O_2 \rightarrow Fe_2O_3 + 2H^+$$

The reaction is quite fast in alkaline to neutral conditions, and quite slow at lower pH. Pyrites is present in the coal and the pH drops; this tends to make the water dissolve any other minerals in the coal. In the case of coal ash the pH can fall to 2.5. Lakes in a coal mining basin containing pyrites can have an average pH as low as 3.0. To remove the solid material from the water it is often necessary to use flocculating agents. These chemicals overcome the charge on the particles and cause them to coagulate and the particles precipitate or became captured in a filter bed. The water is often then suitable for direct use or reuse. It can either be sent back in another line to the mine, or discharged to a suitable water way or reinjected into a suitable aquifer. Reuse of water will result in the accumulation of salts and the salt balance will need to be maintained by suitable blowdown. If further treatment is needed then there is a range of technologies available, including reverse osmosis, use of ion exchange resins and electrodialysis. The most common water reuse technology now is Reverse Osmosis. Reverse Osmosis is used to separate the dissolved solids from the water by applying a pressure over a semipermeable membrane. The water molecules pass through the membrane while with salt is retained by the membrane. The demineralised water is called the permeate. The water on the supply side of the membrane becomes a concentrated salt water stream. Disposal of this saline water stream can pose difficulties, for underground aquifers can be contaminated if the water is discharged arbitrarily. Advice might need to be sought from groundwater hydrologists and appropriate environmental regulators.

18.1.8 Water losses, savings and minor case studies

In recirculating slurry systems the main water loss is from entrainment. Entrained water is that which stays in the transported material semi-permanently after the recapture of the water to be recycled. There can be some small loss due to evaporation. Pump and pipe leaks are typically less than approximately 2%.

As a generalisation, operators of slurry transport lines can make substantial financial savings by increasing the density of the slurry, even when significant investment is needed in upgraded pumps or other infrastructure. The water auditor deals with plant that is already constructed. Even so we note that generally speaking, it is cheaper to install a dense phase slurry system, with no recirculation, than to construct a water return flow line and use say a medium phase slurry.

Often energy savings can be made by careful observation of plant operaton practices. For example, were the raw material supplied to the slurry line to be shut down would the slurry pumps also be turned off? If not very large energy and thus cost savings could be achieved.

In Australia, there are some interesting slurry transport projects. Magnetite is transported to the coast of Tasmania from Savage River, a distance of some 75km. Another is the Century Zinc project in Queensland, involving a slurry of zinc concentrates transported through a 90 km line to the coast, the line having a special internal coating to prevent wear. In this case the supply of water is a contentious issue whereas water is plentiful in the Tasmanian hinterland. Yet the recovery of the ore from the slurry and disposal of the water is an example of an environmental impact. The sea is discoloured and the whole area around the receiving area is black. These examples thus highlight the important issues of the availability of water, treatment of pipes to extend their life due to abrasive flows and the treatment and discharge of the water when it is recovered from the slurry. Binn (1995) describes research leading from a pilot plant to a full scale dense phase slurry system.

18.2 WATER USE IN DUST CONTROL

18.2.1 Introduction

Whether solid, raw materials are transported by trucks or by road trains, rail, ships, conveyor belt or slurry pipeline, often there are stockpiles at either or both the pick-up and the delivery ends. Conveyor belts often carry exposed 'dry' material. The stockpile and conveyor belt loads can lose material in windy conditions as dust. Airborne dust particles can be both an aesthetic blight on the landscape and contribute to a range of tangible matters including health and safety issues. One way to overcome this is to spray water onto the stockpiles or the conveyor load. The water used can range from potable water (rarely) through brackish to saline or even hyper-saline. The availability of suitable water is often an important issue. The use of water to suppress dust has consequences for the raw material. The use of water can change the properties of the raw material. Sometimes this is of no consequence; drying the material returns it essentially to its pre-wetted condition. Sometimes the value of the raw material is diminished, as the properties of the material are permanently degraded somewhat. Some salts in the correct concentration added to coal can lead to explosions in boilers, exemplifying an extreme consequence of using certain quality water.

In short, we aim to find the balance in water use for dust control that will achieve:

- avoidance of an environmental hazard,
- retention of the use of and asset value in materials stored or transported, and
- minimization of water resource demands

We note that dust control is a highly specialized area. We recommend that water auditors engage the services of specialist consultants in the area. Our treatment contains the minimal detail we believe is necessary for the water auditor to understand the issues and to talk intelligently to a specialist consultant. For further details see Schofield & Shillito (1983) and Western Australia, Environmental Protection Authority (1996).

18.2.2 Theory

18.2.2.1 Particle Size

In general finer particle sizes and more even distributions of particle size have greater potential to form dust or to dust at lower wind speeds compared with larger particles. If it is possible to obtain a uniform size distribution, the resistance to dusting is greater even at dry (<5%) moisture content.

18.2.2.2 Wind and moisture

The potential of a material to release dust is proportional to wind speed to some power law greater than one and approximately two. Moisture content has an effect on the intensity of dusting, with the highest airborne dust concentration at dry (<5%) moisture content and relatively stable conditions with approximately 20% moisture content, depending upon particle size. As a rule of thumb fines require greater moisture content for dust stability at approximately 22 percent and coarser particles and uniform grading allows a reduction to 18 percent. Dusting from very coarse material might be controlled with as little as 16% moisture content by volume. Windbreaks of trees or built structures can be used to reduce wind-speed. Where water is scarce, the water used to water the trees, if necessary, needs to be offset against the water saved in wetting the raw materials to reduce dust. Built windbreaks are expensive. The expense of building them will often outweigh the benefit they provide.

18.2.2.3 Wind spatial variability and watering

Stockpiled raw material often forms a rounded tipped cone-like pile, which we take as our example for the purposes of discussion, recognizing that other configurations are possible too. In dead still conditions, the even application of water over the surface would be expected to provide even water content to the material, even after some time has passed and evaporative losses have occurred. With the onset of wind, the rate of evaporation of water varies circumferentially around the pile. Thus losses of water from the surface of the material due to evaporation vary too. The pattern of equal contours of water loss changes with wind speed. If water is applied equally over the surface, it needs to be applied to suit the locations of maximal water requirement. Water is thus wasted, by over-watering other locations on the pile. We can save water by employing on the stockpiles watering cannon having computer controlled spatial variability of the application rate of water. In Figure 18.2 we show a diagrammatic representation of watering rates of a water cannon at a variety of wind speeds.

In addition to this effect, windbreaks will further modify local evaporation rates and thus the spatial distribution of rates required for water application.

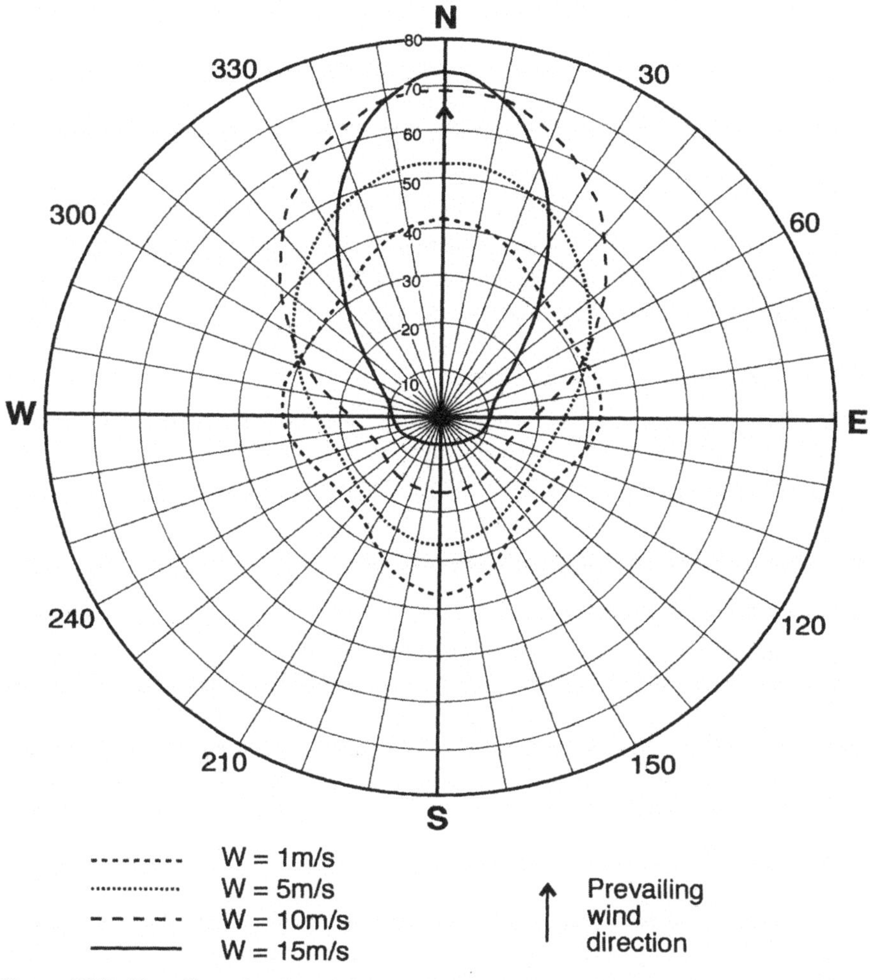

Figure 18.2 : Two dimensional water canon model run at various wind speeds. The contours show the pattern of the water application at different wind speeds, without computer control of the water canon.

18.2.2.4 Evaporation

The relationship between wind speed and evaporation is:

$$E = k(w)\,(V - v)(T - t) \tag{18.1}$$

where E is evaporation rate in mm/day,

$(V\text{-}v)$ is vapour pressure differential between the stockpile surface and the free air stream in hectapascals,

w is wind speed in m/s,

k is a coefficient for the material, and

$(T\text{-}t)$ is the temperature differential between the stockpile surface and the free airstream in degrees Celsius.

Use of this equation requires calibration of the constant against wind speed for a particular material and distribution of material size.

18.2.2.5 Inter-particulate Bonding

Surfactants can be added to the water to reduce the moisture content required to control dust, effectively reducing the value of the constant in equation 18.1. Surfactants are different from the crusting agents referred to later.

18.2.3 Situations requiring dust control

18.2.3.1 Open Ground

Open ground includes areas such as ash dam surfaces, disturbed earth and excavated areas. Whilst some of these can be transitory situations, satisfactorily dealt with by temporary measures such as water trucks, others, such as ash dam surfaces, require more permanent attention. Water is not the only means of dust control. Top dressing with chitter (very coarse material), or crusting with suppressants are options avoiding significant water consumption.

18.2.3.2 Stockpiles

While material passes through stockpiles, the stockpiles themselves are semi-permanent. They can involve large areas, so that careful management of water use can give rise to significant savings of water.

18.2.3.3 Enclosed Transport Mechanisms

Transporting ores by conveyor often involves exposing them to the elements. The most common method of control of dust here is the use of fine sprays designed to impart and maintain the appropriate moisture content.

18.2.4 Water quality, surfactants and crusting agents

If water sprays are used for dust suppression it is rare that freshwater is used because it is expensive. In some cases, use of anything else prejudices the quality of the stored or transported product, so there is no choice other than to use fresh quality water of the least possible quantity. In other cases salt water, crusting agents and surfactants are used to minimize demand on near potable or potable resources.

When dust cannot be controlled using water, it can be controlled using either crusting agents or surfactants. In some mining centers, magnesium chloride is used to crust the surface of ore without penalty. This chemical is a significant component of seawater and, in some parts of the world, of groundwater. The major threat to the break-up of crusts is from personnel and animals walking on them. Surfactants reduce the factor k in equation 18.1 in a different way, reducing evaporation and consequently the water moisture content needed.

Proprietary chemicals can be expensive if used too liberally and should be mixed with fresh water and salt water in an optimum blend. To optimize the blend of suppressant, water and saltwater, a specialist should be approached to identify the optimum point. A brief for a specialist can be prepared if:

- the dust control is a major or significant consumer of water; and
- there is clear opportunity on a cost-benefit basis for reducing water consumption.

18.3 CONCLUDING COMMENT

We reiterate that the control of dust is a specialized area, especially if the water consumption and area involved are large. The Water Auditor needs to understand the issues and identify the opportunities for water conservation or water use efficiency improvements in a preliminary way. S/he should seek the input of a specialist to undertake detailed investigation and evaluation.

18.4 REFERENCES

Binn, T.F. 1995. Progression from research to pilot plant to full size plant dense phase ash slurry conveying of power station ash. 5th International conference on bulk materials, storage handling and transport. 10th – 12th July, Sydney.

British Electricity International 1990-1992 Modern power station practice: incorporating modern power system practice. Pergamon Press, New York.

Jacobs, B. E. A. c1991 *Design of slurry transport systems.* London, New York : Elsevier Applied Science. 312 p.

Schofield, C., Shillito, D, 1983 *Guide to the handling of dusty materials in ports: impact, prevention & control.* British Materials Handling Board, Ascot, Berkshire, UK. 178.

Western Australia, Environmental Protection Authority 1996 *Upgrade of dust management at Finucane Island and Nelson Point, Port Hedland*: BHP Iron Ore Pty. Ltd: report and recommendations of the Environmental Protection Authority. Bulletin 831. Env. Protn. Auth., Perth, W. Australia.

19

Arena: Water use in cooling systems

19.1 INTRODUCTION

Cooling water is used in industry for many reasons. When steam is raised in a boiler for whatever reason, the steam, having been used for its intended purpose, is too valuable a commodity to throw away. The very high quality water used to produce the steam is expensive to replace un-necessarily. It also contains a large quantity of thermal energy, which we wish to regain at least partially. The expended steam is therefore passed through a condenser (see Figure 19.1) to return the steam to the liquid state, ready for re-introducing to the boiler. In the process of condensing the steam, the latent heat of vapourisation is released. Cooling water is the common medium used to receive this rejected heat and to transport it away. In 'once-through' cooling, the cooling water is simply dumped. In recirculating cooling systems the cooling water itself is cooled, often by a cooling tower, before being recirculated to a condenser. The student who wants an elementary account of the theoretical foundations of boilers, condensers, heat exchangers and air conditioners will find useful material in the old but reliable text Rogers & Mayhew (1962) in the chapters 'Flow processes', 'Vapour power cycles' and 'Properties of mixtures'. For an overview of power stations we recommend British Electricity International (1990-1992).

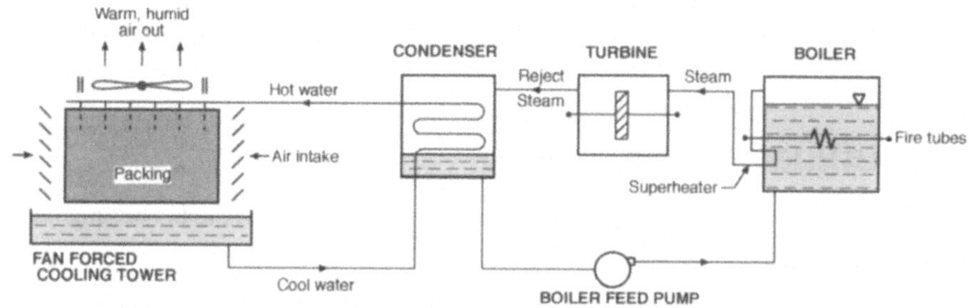

Figure 19.1 An example of a cooling water circuit, where the 'process' for which the steam is used is (arbitrarily) a steam turbine.

A large part of the water that industry uses is cooling water. It is used massively in power generation industries (with some exceptions) and cooling water is needed when energy, carried by steam, is used for industrial processes. Once-through water-cooled power stations receive and discharge huge quantities of water, which is used purely as a medium for carrying heat. More typically, cooling water is lost through evaporation associated with cooling towers, through cooling tower or boiler blowdown (see below) and through liquid (or solid!) water droplets carried away from cooling towers by the drag of either natural or forced air flow. This latter water loss is called drift.

Poor quality cooling water passing through tubes or pipes, through containers or over surfaces, can cause great damage, due to dissolved species or to solids precipitated from the cooling water. The damage can be due to corrosive attack, or due to the deposition of the solids on surfaces. Such solids are called scale. In addition, the thermodynamic properties of the water can be modified in ways that are unhelpful or, sometimes, helpful. Operators of industrial plant requiring cooling water take great efforts to modify the chemistry of the cooling water so as to minimise the harmful properties of the water and to maximise the benign or even helpful properties. This process is summarised under the heading 'water treatment'. The chemistry and operation of cooling water and associated systems can be complex, lying outside the expertise of many water auditors. Generally speaking the water auditor needs a limited knowledge of these matters. If s/he understands the large-scale processes, generally s/he will be able to identify areas for water conservation or areas for improvement of water use efficiency. In this chapter we will introduce various cooling systems to the reader and discuss briefly cooling towers, boiler water system management, water treatment and associated matters.

Given that the water auditor is working on an existing plant, it is most important that s/he understands its operation and can appreciate the importance of the water resource, factors affecting the efficiency of its use and the potential for recovering usable water from wastewater. As is the case in other arenas, the development of an accurate water flow diagram is crucial, whether supplied by the auditee enterprise and updated by the auditor or assembled wholly by the auditor from field observations and measurements.

19.2 COOLING SYSTEM TYPES

Broadly speaking there are two cooling system types, once-through cooling and recycled cooling water systems.

19.2.1 Once–through

In large scale power generation the system most used for cooling is 'once-through'. Essentially this means that a relatively large water body, say a natural or dammed lake, even the ocean, is used both to source cooling water and, when the water has done its cooling, to receive the warm discharged water. In order for once-through cooling to work, the inlet from which the cool cooling water is received must be far removed from the outlet from which the warm cooling water is discharged. Otherwise there could be a gradual increase in the temperature of the water at the receiving inlet, resulting in lower heat transfer in the power station or other industrial process. Typically the inlet water would be drawn from lower in the water column than the discharge outlet. Evaporation is the principal mechanism for loss of heat from the reservoir from which the water is drawn, yet convection and radiation heat losses can be significant.

19.2.2 Recycled

In recycled cooling water systems the cooling water is itself cooled in part of the cooling water circuit. The usual method of cooling is to employ evaporation of a portion of the cooling water in a cooling tower. This process still conserves the bulk of the cooling water because the latent heat of evaporation is very large, allowing sufficient cooling to take place with a small evaporative loss of the water. The evaporation of the water leads to a gradual increase in the dissolved species in the water unless we intervene. The usual interventions are both to implement chemical treatment of the water and to remove some of the water from the circuit and to replace it with high quality make-up water to retain an acceptable water quality in the circuit. Removal of some of the water and its replacement with higher quality water is called blowdown. The potential build-up of salts and the interventions will be discussed more fully below. The water that is discharged from the cooling water circuit can have detrimental environmental impacts and its appropriate disposal or treatment is a significant matter. Sometimes the discharged water can be used in lower order water quality streams in an associated industrial process. Recycled cooling water systems can themselves be subdivided into different types.

19.2.3 Wet cooling systems

Wet recycling cooling water systems use a cooling tower to enhance the evaporative extraction of heat from the cooling circuit water. There are two general types of cooling tower, natural updraught and mechanical fan-forced. Natural updraft cooling towers have a very high capital cost, which is reduced considerably by using the much more compact fan forced cooling towers. The advantage of low capital cost is then accompanied by the disadvantage of high operating costs.

19.2.4 Dry Cooling Systems

Strictly, dry cooling systems do not belong to the category of recycled cooling water systems, but we include this discussion here in preparation for the section below. When water costs are very high dry cooling systems can be installed and a number of power stations in the US and Europe are using wind to directly cool boiler water through radiator type cooling towers. These are very expensive, as the contact area required is very large. Yet there is no requirement for making up cooling water losses and demineralised make-up water is only required for the boiler system itself. Dry cooling is applicable in areas of low and expensive water resources and high wind.

19.2.5 Wet/dry Cooling

'Wet/dry' cooling amounts to an enhancement of the dry cooling system. This is achieved by spraying water onto contact surfaces of the 'radiator' and increasing the normal heat losses (by convection and radiation) with evaporative heat losses from the water on the surface. The size of the radiator can be significantly reduced compared with purely dry cooling. The water sprayed onto the radiator surface is part of a recirculating cooling water system and, because of the evaporation, the circuit must be blown down. Blowdown water treatment is possibly required. The major benefit of wet/dry cooling systems is to reduce water demand. Energy usage is less than wet systems but the capital cost is higher. Wet/dry cooling is not recommended for humid locations.

19.3 COOLING TOWERS

Cooling towers are important parts of most cooling water systems and they warrant further discussion.

19.3.1 Natural Updraught Cooling Towers

A natural updraft cooling-tower consists of a very large, semi-hollow, hyperbolic paraboloid structure, typically 120 m high. Figure 19.2 illustrates such towers at the Bayswater power station in New South Wales. The structure generates a natural updraft of air drawn through openings in the bottom, up the interior of the tower through packing or a contact medium, into a faster moving portion of the planetary boundary layer at the tower top. Water jets or sprays supply hot cooling water from the condenser high up in the tower, from whence it trickles down over packing to a large basin at the tower bottom. On the way evaporation, accompanied by convection and radiation, cools the water, which it is pumped back to the condenser from the basin. Water savings in a natural updraught cooling tower can be achieved by reducing drift with splitter boxes, replacing damaged packing material, and looking for leaks in the pipes, pumps and basin. Leaks in condenser tubes normally show up by the detection of increased conductivity in the boiler water system. A discussion on water losses due to blowdown will be deferred until later. The natural updraught cooling-tower has drift losses of approximately 0.04% of the flow rate in the cooling water circuit.

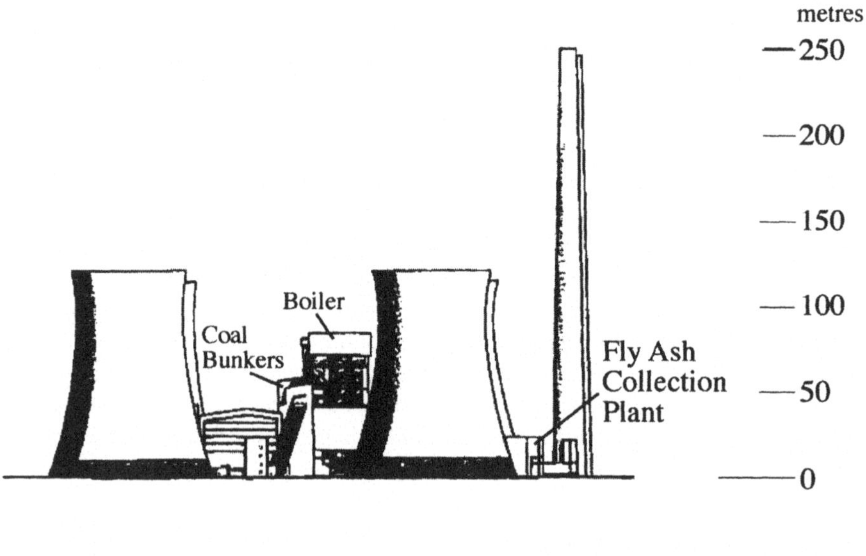

Figure 19.2 Pictorial representation of natural updraft cooling towers in a power station. (Source: John Schlafrig)

19.3.2 Mechanically forced cooling towers

Mechanically forced cooling towers are installed in wet cooling water circuits where the costing is favourable and also where the height of a natural updraft cooling tower is unacceptable. The cooling water is sprayed into the air inside the tower and the air is driven horizontally or vertically by large fans to cool the water as it trickles down the packing and returns to the cooling tower collection basin, from whence it is pumped back to the condenser. Normal drift losses are of the order of 0.25 to 0.50% of cooling water flow rate. Because the drift is in liquid (or solid!) phase rather than in vapour phase, it carries away small quantities of impurities. The advantage of drift is that the need for blowdown is reduced, because of the removal from the cooling water of impurities in the drift. The disadvantage is that drift reduces the volume of water in the cooling water circuit and this reduction must be made up with appropriate quality water. Water savings can be made in the mechanical cooling tower area by replacing damaged packing and checking each of the performance of the fan, spillage of the basin and leaks in the pipes, pumps and basin.

19.3.3 Some Details of Cooling Tower Operation

When cooling water is evaporated in a cooling tower the dissolved and suspended solids present in that water remain. Because dissolved salt has a maximum solubility limit, deposition occurs when concentration of the salt in the cooling water is permitted to continue unchecked. The salts of lowest solubility, such as calcium and magnesium carbonates and silicates, precipitate first. Any suspended matter in the tower will concentrate in a manner similar to that of dissolved salts. *Notionally* the water recycles through the cooling water system a number of times before concentration of impurities rises to the limit of acceptability and requires the whole of the cooling water to be discharged and replaced with treated new water (see below) to retain an acceptable cooling water quality. The number of cycles in this *notional* process is called 'cycles of concentration' (COC).

We hasten to add that in practice the cooling tower is *not* operated in this way, for water treatment and blowdown occur periodically, or even continuously, to maintain the water quality in the cooling water circuit at an almost constant acceptable level. Nevertheless the concept of cycles of concentration is useful. It can be thought of as the mean residence time of water in the cooling circuit using units of 'cycles' (or 'passes') through the cooling water circuit. Another way of thinking of cycles of concentration is the ratio between the concentration of total dissolved solids (TDS) in the cooling tower water and the concentration of TDS in the system's make-up water supply. For example, cooling tower water containing four times as much TDS as its make-up supply would be operating at four cycles of concentration. These two ways of thinking about COC can be shown to be equivalent. The cycles of concentration may be maximised by water treatment or better quality make-up-water or both. This is highly desirable from a water auditor's point of view, contributing significantly to water conservation and water use efficiency. Note that make-up water must be added to compensate for that lost by evaporation and drift as well as from blowdown.

We can relate blowdown to water lost by drift (or windage), evaporation and to the cycles of concentration. A mass balance of water in the cooling circuit leads to

$$B=\frac{E}{C-1}-D, \tag{19.1}$$

where B is blowdown as a percentage of total water circuit water mass,
E is evaporation as a percentage of total water circuit water mass,
C is cycles of concentration,
and D is drift as a percentage of total water circuit water mass.

The optimum cycles of concentration for a cooling tower are governed by each of the following - tower design, the cooling water quality and treatment regime and the circuit operating conditions. We suggest that the equipment manufacturers' advice is sought.

19.4 COOLING WATER QUALITY AND DISCHARGE ISSUES

From equation 19.1 it is quite clear that a high cycles of concentration results in minimising blowdown and thus water is conserved. In order to attain high cycles of concentration the quality of the water must be very good, either because of its innate quality prior to introduction to the cooling water circuit, or because it is made 'high' quality by treatment. Chemical treatment is an expensive process and thus the correct treatment, which neither under-modifies the water nor overuses chemicals, is important to achieve. Such an achievement will minimise plant downtime and maximise plant life. In particular water quality affects

- tube life in the condensers,
- cooling tower life by minimising concrete attack where concrete is used,
- pump life,
- pipeline life and,
- hydraulic characteristics.

Scaling is a well-known problem arising out of the concentration of dissolved solids in cooling water systems. A common measure of the adequacy of water to avoid scaling is the Langelier index, which should be less than one. Calcium and sulphate can pose particular problems, but generally the salinity of the cooling water circuit makeup water determines the cycles of concentration. It is therefore in the operator's interests to use make-up water very low in calcium, sulphates, and salinity generally and, because of the effects of suspended solids on the efficiency of heat transfer surfaces, low in particulate matter and organic matter.

19.4.1 Blowdown Treatment

The blowdown water can be discharged to water courses if it is of an environmentally acceptable quality (as determined by environmental regulations or agencies). Alternatively, the blowdown water can be used on site for a variety of 'lower order' tasks, eg. washdown of vehicles. Or it is often cost-effective to treat the blowdown water to concentrate the impurities and to reuse the cleaned water either in the cooling system or elsewhere. Membrane concentration such as reverse osmosis (RO) is one such possible process, others being use of ion exchange resins, electrodialysis reversal (EDR) and vapour distillation. Distillation of the blowdown water is a very energy intensive process, which would normally be employed only where large quantities of waste heat are available to be diverted to the distillation process. Therefore blowdown treatment normally involves membrane concentration such as reverse osmosis (RO) or electrodialysis reversal (EDR), ahead of vapour distillation. Lime softening can be employed with hard water. Note that RO and EDR recover up to 80% of feedwater and produce concentrate about five times the original salinity. Distillation will recover about 97% of water and will produce concentrate approximately 35 times the original salinity.

19.4.2 Concentrate Disposal

Savings can be made by recovering water from blowdown using a range of options and disposing of the concentrate. Further efficiency gains may be possible by further concentrating the concentrate into crystallised salt and recovering the water. One or more of the following methods normally disposes of the concentrate from blowdown treatment: crystallisers and landfill, pumping to the ocean, conditioning of ash or tailings and salt recovery. In order to reduce handling costs it could be cost-effective to install a brine concentrator.

19.5 BOILER WATER CIRCUIT

19.5.1 Boiler Make–Up

The quantity of water required for boiler make-up is much less than that needed for cooling water make-up. In approximate terms water use is 4kL/day/mW of boiler rating, varying somewhat with boiler age, and tube material. Although this is not a large volume, the quality of the make-up water needs to be very high and it rises with boiler pressure. One major cause for blowdown of the boiler circuit, and thus the need for makeup with high quality water, is leakage of lower quality cooling circuit water into the boiler circuit, often from the condenser. Such leakage is normally detected by a conductivity meter probe in the boiler and possibly in the steam feed system. The high quality of make-up water to the boiler is achieved through a demineralising plant. Commonly degasification, softening and ion exchange are employed.

19.5.2 Condensate Polishing

Where minor water or steam quality worsening occurs in the boiler steam cycle, a process often used is condensate polishing, which is a pressurised ion exchange system removing impurities from partially contaminated steam ready for boiler water make-up from the de-mineraliser plant. It is minor cleaning of boiler circuit steam without major pressure loss.

19.6 POWER GENERATION

Power generation accounts for the highest volume of cooling water usage in cooling water systems because of the large water needs for condensers. This is true whether the power station is fossil fuel steam plant, diesel or nuclear. Of the newer technologies for generating electrical power, gas turbines are probably the most common world wide and use little water. Technologies using renewable energy often use little water too. For example, around 50% of Denmark's electricity is generated by wind turbine. Ocean thermal energy conversion (OTEC) will be a significant user of water, while photo-voltaics use none. There is a strong likelihood that in the longer term future a hydrogen economy might come into being. If electrolysis of water becomes the preferred means of generating hydrogen, large quantities of water will be required. However large quantities of warm water will not be discharged to the environment.

It is still true that the majority of the world's electricity is generated by coal or other fossil fuelled steam generating plants. Thermal efficiency of coal-fired plant has recently exceeded 40% and is improving rapidly with the improvement of materials for high temperature and pressure environments. Even so, huge quantities of energy leave these plants as warm cooling water. It is either dissipated as low grade energy via cooling towers, or discharged into large water bodies in once-through systems, with the consequent risks to local ecosystems. We note the following rules of thumb. Both natural updraft and mechanically forced cooling systems use approximately the same volume of circulating flow of approximately 25 L/s per Megawatt. Of this, typically 34 kL/day is lost in evaporation to the atmosphere and 2 kL/day is lost in drift. From the water auditor's point of view, a shift to electricity generating plant using renewable energy will be welcome.

19.7 FURTHER PROCESSES

19.7.1 Air Conditioning

Heat pump air conditioning and heating has higher thermal efficiency than evaporative air conditioning. However, the capital costs are much higher for heat pumps than for evaporative air conditioners, and the balance between capital and operating costs is dependent on energy and water tariffs. In dry climates, evaporative air conditioners have their place, but they use significant volumes of water. If not maintained correctly, they will use excessive water and deplete water quality in the unit to the point where the plant life reduces by several years. Evaporative air conditioning is very similar to a cooling water system for power generation, or process use, except that it works in reverse. For a known volume of evaporated water, air temperature can be reduced by a known amount given the psychrometric conditions of the air. Water evaporates while running down packing screens through which air is drawn by a fan. This process leaves behind impurities, which need to be blown down. Drift is high in evaporative air conditioners, but many of the impurities deposit on the packing and require flushing either as blowdown, or by compressed air once a year in a standard service of the plant. Blowdown rates for evaporative air conditioners are normally proportional to the rating of the unit (kW). In modern units the blowdown rate is often controlled automatically by an electronic device. To save water, the manufacturer and the water auditor can review the flow rates and set up annual servicing of core material.

19.7.2 Others

Steam and heat derived from steam are required in a myriad of other processes than power generation. For example steel and aluminium manufacture and sheet production use large quantities of water, especially for cooling. There are also many other industrial processes that require both steam and cooling water, such as the bottled beverage industry, the food and the paper industries. The metal industries lie outside the scope of this book, but the latter industries are covered in separate sections.

19.8 REFERENCES

British Electricity International 1990-1992 *Modern power station practice: incorporating modern power system practice*. Pergamon Press, New York.

Rogers, G.F.C. & Mayhew, Y.R. 1962 *Engineering thermodynamics, work and heat transfer*. Longmans, London.

20

Water in the non-metal, non-food industries – the paper industry

20.1 INTRODUCTION

Non-metal and non-food industries cover a broad spectrum of activities, and many different water use requirements. The applications for water use in the chemical industry can be different to those requirements of the fibre box industry and different again for the textile industry. This chapter will cover the applications of water in the pulp and paper industry, which, because of its requirements of scale and quality, combines some principles applicable to other industries.

The pulp and paper industry is based on the properties of the cellulose fibre. These fibres occur in a variety of plants and a number have been, and still are, used as feedstock for pulp mills. The most common source of fibre is wood, although straw, cotton, hemp, bagasse etc. can also be used. The cellulose fibre has some unusual properties, when dispersed in water, that make it suitable for the manufacture of paper. The outer layer of the structure swells and almost dissolves in the surrounding water. The water and the cellulose molecule exhibit a property known as hydrogen bonding. When the surrounding water is removed by drainage, pressing and drying in paper making, the cellulose molecules will then hydrogen bond with anything else suitable, usually other cellulose molecules on other fibres. The hydrogen bonding is therefore the "glue" that holds the fibres together in a sheet of paper and why water is such an essential element in paper manufacture.

20.2 FIBRE PREPARATION

The usual source of cellulose fibre for paper manufacture is wood and there are a variety of methods used to produce pulp (fibre) from wood chips. In wood, the cellulose fibres are 'glued' together by a natural polymer called lignin; pulping frees the fibres so that they can be suspended in water for paper making. The pulping method, and the type of trees selected, usually depend on the type of paper that is to be made. The pulp is therefore classified according to the source of the fibre and the method used to pulp it. Coniferous trees normally produce longer fibre pulps (or softwood), and eucalypt fibre is classed as short fibre (or hardwood). A variety of pulping methods can be used, either mechanical, chemical or a combination of both.

This presentation will concentrate on one type of pulping as an example of a pulp mill water circuit. A summary of the Kraft process for producing chemical pulp will be presented here. The actual process is considerably more complex but examples of the main water applications will be discussed. An example of a waste paper recycling plant is also presented.

20.3 KRAFT PROCESS

20.3.1 Process

The Kraft process dissolves away the lignin component of the wood using caustic soda catalysed by sodium sulphide (see Figure 20.1). Chemicals are added to the wood chips in a digester where the period of time at a given temperature determines the extent of cooking (lignin removal). The mixture is then discharged at pressure from the digester into a "blow tank" and the dissolved lignin (called black liquor) is then separated from the pulp by washing in a series of counter current washers. There are different types of washing systems but drum type washers will be presented here.

The pulp and black liquor is added to the first washer where it forms a mat on the rotating first washer drum. Liquor is drawn out of the mat by vacuum and cleaner water is showered onto the mat to displace the concentrated liquor.

Extracted liquor is separated from air in a sealed tank below the washer. Water for the first washer showers is drawn from the second washer seal tank. The mat from the first washer is repulped in water from the second washer seal tank and then forms a mat on the second washer drum. This process is repeated a number of times (usually three or more) and fresh or recycled (clean) water is added onto the final stage washer showers. This is a counter-current washing cycle.

The cleaned pulp is usually stored after washing for use directly on the paper machine for brown grades or pumped to a bleach plant if white paper grades are required.

Black liquor from the pulping process is pumped to the chemical recovery section of the mill. Chemical recovery will remove the lignin component of the black liquor and regenerate the cooking liquor (caustic and sulphide). The black liquor is first evaporated in multiple effect evaporators to raise the solids content above 50%. Condensate is recycled to the pulp mill as wash water on the last washer. The black liquor, when sufficiently concentrated, is injected into a 'Recovery Boiler'. This is a furnace that incinerates the organic component of the black liquor (using it as a fuel) and generating steam (in the tubes that line the walls of the furnace). Sufficient steam is produced to generate an excess of electricity and process steam. The inorganic component of the black liquor is removed from the furnace as molten salts, which are then dissolved in water. This solution is causticized with slaked lime and clarified to make cooking liquor again.

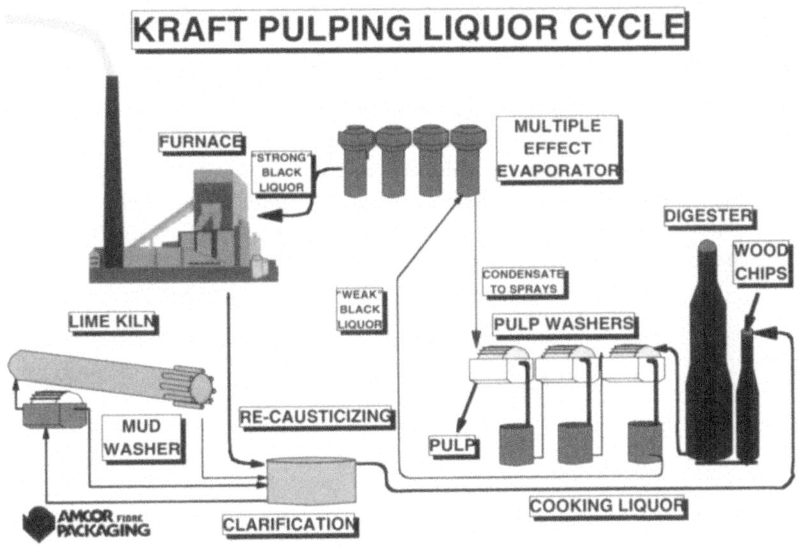

Figure 20.1. Kraft pulping liquor cycle (water cycle). (Courtesy of Amcor Fibre Packaging).

20.3.2 Water Circuit

Evaporator condensate and makeup water enters the pulp mill water circuit at the last washer showers and displaces black liquor from the pulp mat (on the drum) into the filtrate. In order to wash as much black liquor as possible from the mat an excess of wash water is required, but since this water must later be evaporated a balance between solids removal and evaporation cost must be optimised. As described above in a counter-current washing circuit, the filtrate from each washer is used as wash water on the preceding washer until the black liquor reaches the storage tank used to supply the evaporator set. The evaporators can return over 70% of the water to the showers when noxious substances are removed from the condensate. The balance of the water goes forward to the Recovery Furnace and is evaporated and lost to atmosphere as the concentrated Black Liquor is burned. The inorganic salts are retained and are dissolved again in water for recausticizing. The water then returns to the pulping circuit in the pulping chemicals. This circuit cannot be completely closed as salts extracted from additives and raw materials (i.e. the wood) will accumulate if not bled from the circuit. The presence of high levels of some ions (i.e. chloride) will increase corrosion rates and accumulation of other ions (i.e. metals) will affect the performance of the recovery process.

20.4 BLEACHING

The bleach plant can be the largest user of water in a pulp mill as it requires high quality water (that will not cause a reduction in brightness of the bleached pulp) and it is usually impossible or not economic to clean and totally recycle the filtrate. This filtrate is usually coloured and contains significant amounts of chloride and chlorinated compounds, as many plants use chlorine based bleaches. The bleaching circuit must have a significant effluent flow to discharge the dissolved material that would otherwise accumulate.

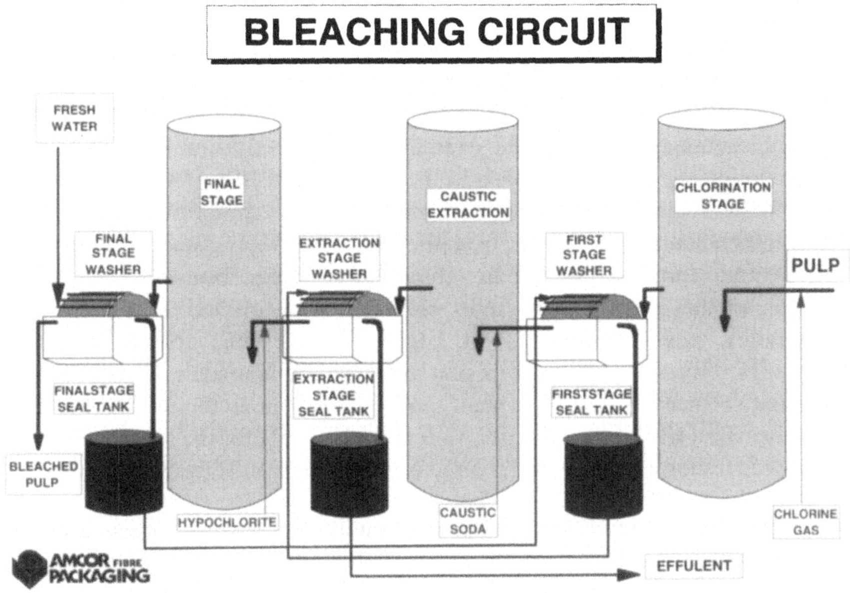

Figure 20.2 A typical bleaching circuit. (Courtesy of Amcor Fibre Packaging).

20.4.1 Process

The bleach plant is usually a multistage process involving the injection of the bleach through a mixer into the pulp, then a retention vessel which holds the conditions constant for the reaction time required by the bleach, and then usually (but not always) followed by a washer to remove spent chemical and residues extracted (see Figure 20.2). This is repeated for subsequent stages. The bleaching sequence in the example is a relatively simple system and much more complex sequences are used in practice. A number of factors are considered when selecting a bleaching sequence. Some of these are capital costs, chemical costs, limits on effluent quality, input pulp quality, bleached pulp quality requirements, water quality and availability.

A significant development effort is being directed at closing bleaching circuits in recent years, mainly by substituting the chlorine based bleach chemicals by other chemicals. Some new plants use oxygen as a first stage bleach, and others have been built to use ozone. Other work has been directed at using modified evaporators to recover water for reuse and to concentrate soluble rejects for disposal.

20.4.2 Water Circuit

The fresh water is applied to the last washer to displace final stage residuals and the filtrate from this stage will be used in one of the preceding stages. The where this filtrate is used is dictated by the design of the bleach plant, chemicals used, etc. The bleaching sequence in the example is a relatively simple system, but it is worth noting the concurrent wash in the first two washers. This is possible as most coloured material is not released from the pulp until the caustic extraction stage and the alkaline wash water from the final hypochlorite stage also provides some neutralisation of the acidic chlorination stage. Because the caustic extraction washer filtrate is the most coloured, it is rejected as effluent. The concentration of contaminants (both organic and residual chemicals), water quality, pH, type of bleach, corrosion potential, temperature, bleached pulp quality requirements, and bleach plant sequence are among the factors that need to be considered when reviewing the water requirements for the bleach plant.

The preparation of bleaching chemicals is not shown in the sketches but all water used in these plants for chlorine cell plants, scrubbers, chemical dilution is high quality fresh water. There is little opportunity to use lower grade water in such applications as the contaminants would consume some of the bleach chemical and/or interfere with the production of chemical. Cooling water and other similar applications can use lower grade water.

20.5 RECYCLED PAPER

Pulp suitable for manufacture of many paper grades (white or brown) can be sourced from waste paper. The waste paper is normally collected from offices, shops, households, printers, etc. and hence can contain a variety of paper types and contaminants depending on the source. The fibre quality is therefore variable unless the input paper is sorted in some way. The cleanliness requirements of the pulp produced from waste will determine the selection of waste paper type (as feedstock) and the type of cleaning required in the waste paper plant. The cleaned pulp from a waste paper plant can be either made into paper without further processing or it can be bleached.

20.5.1 Process

A waste paper plant is designed to remove contaminants (predominantly visible ones) from the fibre and will normally consist of a pulper to disperse the fibre in water then successive stages of screening and cleaning equipment to remove contaminant materials (see Figure 20.3). Some screens are designed to reject particles on the basis of shape and others (e.g. hydrocyclones) reject particles based on density differences. Chemical separation can also be used; chemicals can be added to assist the disintegration of the pulp or they may be surfactants that attach to ink particles and, with air injection, float off the ink and other contaminants. Dispersion with steam is also used to break up particles so large particles are not readily noticeable in the final sheet of paper.

The example given in the flowchart could be for a plant used to repulp used (old) corrugated containers (OCC). The waste paper is normally repulped to at least 3% consistency (3 g dry fibre in 100 mL water) and the initial (coarse) screening is done at this consistency, removing large contaminants (e.g. nails, staples, rocks, plastic bags) but fine screening is often done at less than 2% consistency and removes smaller particles (i.e. sand, grit, small pieces of plastic). Both screens that reject by 'shape and size' and hydrocyclones that reject on 'density difference' would be used in both the fine and coarse sections of the plant. The accepted fibre requires a thickening stage following the fine screens to raise the consistency to over 25% for feed to the dispersion plant. Water is extracted using a thickener, which is similar to the drum washer above but without wash water sprays, and then followed by a screw press or presses to achieve the consistency aimed for. The pulp would then be heated to a high temperature and mixed to disintegrate and disperse any remaining contaminants (i.e. waxes or some adhesives). Dilution water is added after dispersion to lower the consistency to about 4% so that the pulp can be pumped to the paper machine and/or proportioned to other additives or pulp types being used.

20.5.2 Water Circuit

The water extracted from the pulp at the thickening stage would be used for dilution of incoming paper in the pulper and dilution throughout the circuit (i.e. ahead of the fine screens). Because the paper entering the pulper is dry (~ 90% consistency) but leaves the circuit at about 4% consistency, there is a net circuit imbalance which is normally satisfied by makeup by excess water (backwater) from the paper machine circuit.

The waste paper plant is not designed to remove dissolved materials, i.e. starch adhesive used in corrugated boxes, salts in supply water and chemicals used in pulping; therefore these will accumulate in the circuit until discharged either in the paper being manufactured, or in the effluent stream. It is often considered more financially attractive to design the total water circuit (paper machine and waste paper plant) to minimise the amount of fibre and dissolved material being lost in the effluent stream, but this should be carefully analysed in each situation. There can be significant operating costs involved in running the circuit with a high circulating load of dissolved material.

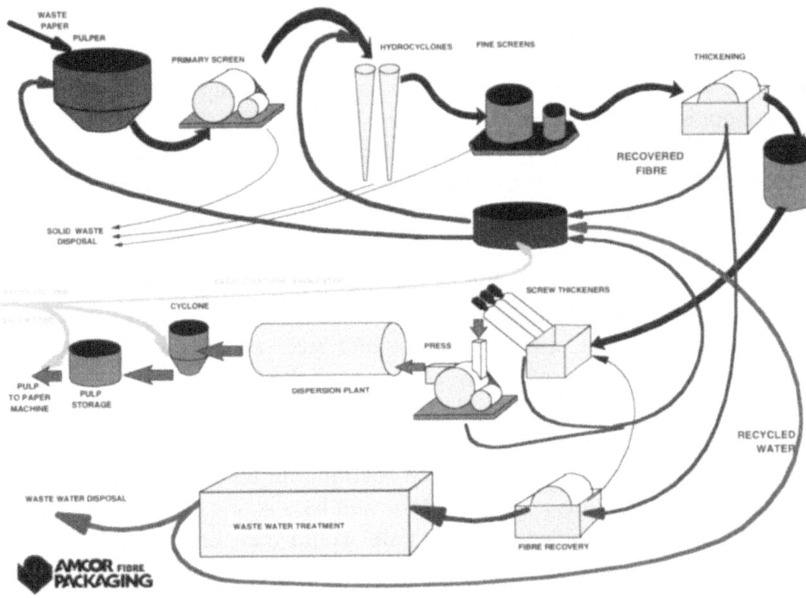

Figure 20.3. Waste paper plant water circuit. (Courtesy of Amcor Fibre Packaging).

20.6 PAPER MANUFACTURE

20.6.1 Process

Pulp is pumped to the paper machine and usually diluted to 1% or less before being drained on the machine wire (a plastic mesh) (see Figure 20.4). Low consistency is necessary so that fibres are uniformly dispersed. If the consistency is too high the paper produced will have poor formation (i.e. will be 'lumpy'). In the initial stages of formation, water will readily drain from the pulp but vacuum assisted water extraction is required as the pulp becomes a wet sheet of paper. The paper, now self supporting, can be removed from the wire mesh on which it was formed. Further water is removed by pressing in several stages which raise the consistency to about 50%. The water pressed from the paper is transferred into the felts which run through the press against the paper. This water is then removed from the felts by vacuum. The pressed paper sheet is dried by direct contact with steam heated drying cylinders and hot air (the dryer section). Some paper machines have a size press or coater installed in the dryer section which adds more water with additives (ie starch to increase paper strength properties) which also must be dried. The final moisture in the finished product is usually about 7% (but depends on quality requirements).

20.6.2 Water Circuit

Water that is drained through the wire during paper making is collected in a backwater tank and is used to dilute the incoming pulp. It may also be filtered and recycled for use in sprays. The water extracted by vacuum and into press felts can be filtered and reused or pumped to the backwater system. The water contained in the wet paper entering the dryer section is evaporated and hence is lost to the water circuit. Water has many applications around the paper machine and recycled water can be used for most of these provided it can be economically cleaned to the required quality. It is used to clean machine wires and press felts, provide sealing in liquid ring vacuum pumps, lubricate felts, transport medium for the pulp. The quality of water used in most applications has a very significant effect on the machine production rate and product quality. Sprays require water with low or manageable levels of suspended solids, to avoid plugging but the control of corrosion will require management of BOD and dissolved salts levels in the overall circuit. The sketch shows a hypothetical paper machine 'wet end' water circuit (Figure 20.4). Sprays are located on each item of the machine 'clothing' (the felts and wires) and are used to lubricate and clean these fabrics while the machine is in operation. The requirements of each application determine the quality of water required. Some of these applications are discussed below.

20.6.3 Sprays

Sprays in the press section will require water with very low suspended solids, but other applications can tolerate a higher level of suspended solids. Other water quality factors will need consideration when assessing water quality requirements for sprays, but they are usually less important than the content of suspended solids. Filtration should be provided ahead of all machine sprays to eliminate blockages. The type of filtration installed will be determined by the quality of the feed water, desired 'accept' water quality, cleaning cycle or filter element replacement frequency and economics. If a spray is normally operating on high quality (i.e. town water) and protective filtration is required to prevent (occasional) spray blockage, then the most appropriate system will be a manually cleaned or replaced cartridge filter. There will be long time intervals between servicing the filter, therefore the operating cost will be relatively low. A filter system for treating water containing high suspended solids with a broad spectrum of size will need automatic cleaning to be economic, as the time between cleaning cycles will be short and hence too frequent for a manually cleaned system. Selection of filtration equipment for the paper industry usually involves a number of trials before a suitable system is identified. Fibre characteristics (fibre stapling of wires etc.), suspended solids concentration and size distribution are among the factors that need to be considered, followed by a detailed financial analysis to compare the various options. For example it may be uneconomic to recover machine backwater of high suspended solids levels for sprays if cleaner water of moderate cost is also available and the disposal cost of the extra volume is minimal.

20.6.4 Vacuum Pump Sealing

The seal water should not contain any dissolved materials near the saturation limit as some water will evaporate in the pump as the pressure is reduced, increasing the concentration above saturation, causing precipitation of solids in the pump interior, reducing the clearances and eventually binding the rotor and casing. The acceptable concentration of dissolved material is a function of the nature of the solids, the temperature, the operating vacuum, pH and the efficiency of any dispersants added.

20.6.5 Dissolved Solids in a Mill Water Circuit

The water system dissolved and colloidal solids concentration (generally higher when using waste based pulp) can affect the performance of some additives used. High concentration of colloidal solids can reduce the effectiveness of drainage or retention aids. Starch and/or many other additives may be required to achieve the required product quality; while high dissolved solids concentration can reduce additive efficiency and increase production costs. High concentrations of some inorganic salts (e.g. chloride) will also significantly increase corrosion rates. Paper machine designs, the types of products made and the raw materials used show considerable variation, and therefore it is essential that each system is properly understood before credible recommendations and financial analyses can be made.

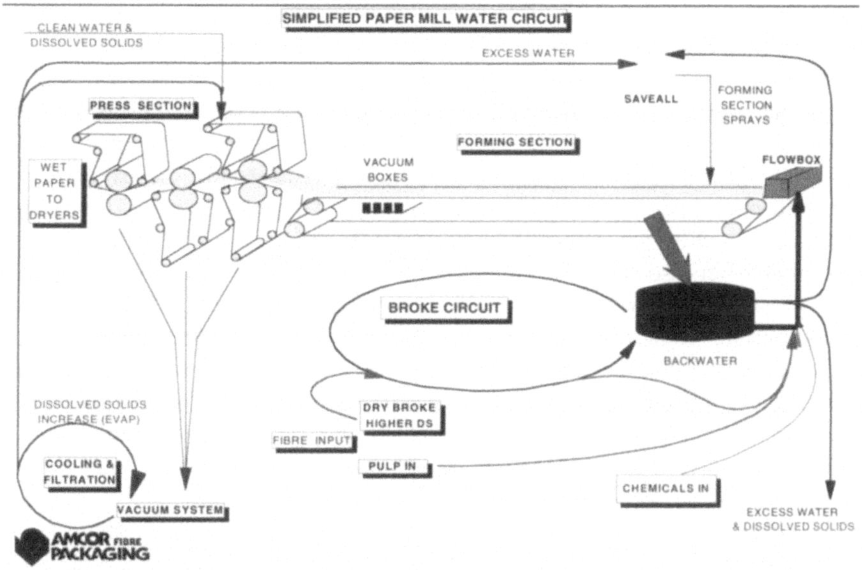

Figure 20.4 Simplified paper mill water circuit. (Courtesy of Amcor Fibre Packaging).

20.7 EFFLUENT TREATMENT AND TREATMENT FOR RECYCLING OF WATER

20.7.1 Process

Options for water reuse fall into two broad categories; an internal recycle loop in which water is recycled back into the same or an adjacent process in the overall process flow; or a longer (usually external) recycle loop that involves taking water from one part of the process, then returning it to a different part of the process circuit. Examples of the internal recycle loop would be: when paper machine backwater is filtered and returned to the sprays on the wire (the water stays within the forming process loop), or in a pulp mill washer, when filtrate is recycled from one washer and is delivered to the showers on the previous washer (the water recycles to an adjacent washing process.)

Examples of the external recycle loop would be: removal of water from a Waste Paper Plant biological treatment (to remove dissolved organic material), clarification, filtration and then returning the water to machine felt sprays; or taking black liquor (filtrate from the first washer in the pulping process) evaporation to remove the water and return of this water (evaporator condensate) to the last washer. The water is returned to a different part of the process in both cases and its properties have been significantly modified.

Reuse of water in an internal circuit is usually the most cost effective solution to reducing overall water consumption. The physical and chemical properties of the recycled water and those of the water already circulating in the process are usually similar, and this minimises the treatment required. In the example of paper mill backwater, water is recycled to sprays. Therefore the only treatment required is usually the removal of suspended solids (e.g. filtration, or air flotation clarifiers) so spray blockage does not occur. More expensive processes such as evaporation, which would also remove dissolved solids, are therefore not appropriate in such an application. However careful consideration is still required to ensure that the quality requirements of the recycled water match with those of the application and the process it is entering. The cost of improving water quality is related to the volume treated and the type and degree of the treatment required. It is therefore rarely economic for water quality to significantly exceed the quality requirements of the application.

The water treatment systems in external recycle loops are often more sophisticated and expensive than those used for internal loops and similar to those used for effluent treatment. There is usually also a major change in one or more water quality parameters, e.g. the removal of dissolved organics in a bioreactor, or dissolved inorganic salts in an evaporator. In these examples, external treatment is used to remove or control the concentration of a contaminant that would otherwise be detrimental to the total process (or receiving environment in the effluent treatment case). When extensive water recycling (usually internal) is practiced, the concentrations of the contaminants, not removed by internal treatments, can accumulate to unacceptable levels. More complex (external) water or effluent treatment will be required before further water recycling can be considered.

There can be some cases when recycling through an external loop can be more economic than the internal recycling loops and conversely, the internal treatment alone may be adequate for waste water to meet required effluent quality standards. This must be considered when assessing a project.

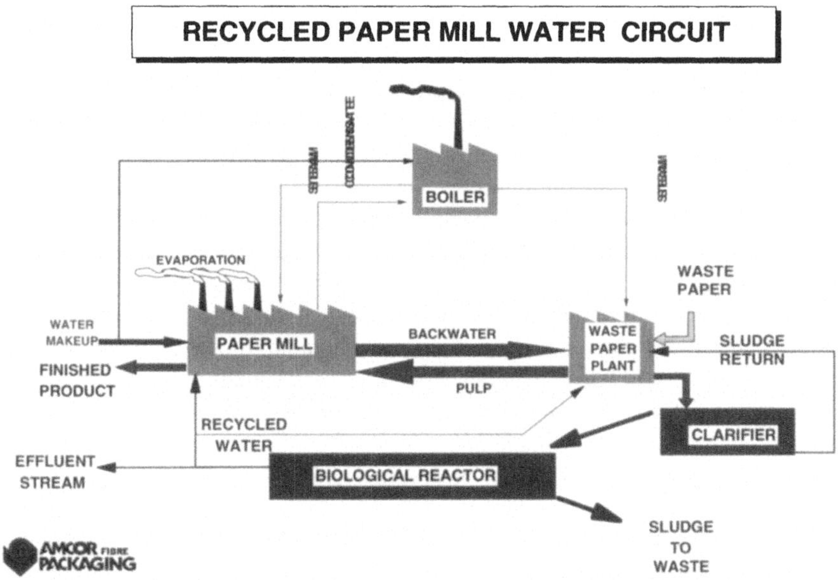

Figure 20.5 A typical recycled paper mill water circuit. (Courtesy of Amcor Fibre Packaging).

20.7.2 System Design / Equipment Selection

A proposal to reduce total process water consumption (or effluent treatment), should first consider the current mass balances of suspended, dissolved organic solids (COD), dissolved inorganic solids and any specific material identified as problematic. The maximum allowable levels for all these materials must then be established. Each potential application for recycled water (or outlet for effluent) must then be considered and the quality parameters defined. Systems that can achieve the required water quality can then be identified, the balances for all the above species calculated and the effect of the water recycling on the total process estimated. For example, the recycling of treated mill effluent will affect the levels of dissolved solids, corrosion, temperature, suspended and colloidal solids, etc, in the mill circuit, and may affect the operation of the machine and the quality of products made. These factors must be considered before making such recommendations.

The selection of a design for a water recycling circuit is often an iterative process. A number of alternative systems and different equipment types are compared on the basis of capital, operating cost, reliability and manning requirements. The selected system is usually trialed on a pilot scale, then the balances and economics recalculated to assess the technical and economic viability of the system.

Many systems are used for treating water for effluent discharge. The systems used often reflect the particular quality requirements of the receiving waters or of the municipal sewer. Sedimentation or air flotation clarifiers are commonly used for primary effluent treatment to remove suspended solids. The primary treatment sludge is often returned to the paper mill to recover any fibre lost due to leaks, tank overflows, etc. The next treatment stage is frequently a biological reactor (anaerobic or aerobic) to reduce the BOD content in the final discharge. Some of this water is often returned to the mill provided dissolved solids levels are satisfactory and the remainder is discharged to maintain those dissolved solids at acceptable levels. Discharge volumes from mills are usually large and therefore most techniques for recovering water and concentrating the soluble fraction (e.g. evaporation) are usually uneconomic.

Water acquisition and disposal costs are significant operating costs in the pulp and paper industry, and therefore water efficiency and water costs are usually scrutinised to reduce expenses. The limitation on water efficiency is often related to the need to remove an effluent stream that will contain the contaminants that would otherwise concentrate in the process and create operating problems and costs.

20.8 AUDIT

An Audit of a pulp and/or paper mill's water management will need to include an overview of the current process, operating practices, and identify any unique features. Benchmarking the plant against similar mills may be a useful guide. The Auditor will need to identify the water quality factors (for both the overall system and individual applications) that will influence production rates, manufacturing costs, product quality, equipment reliability, and maintenance costs. Process and water flowcharts need to be provided or developed and used with the water quality data to review the process and identify potential inefficiencies. Preliminary analysis of these areas should identify the scope for improvement and assess the technical and economic feasibility of any changes. The effects on the overall plant must also be assessed. This preliminary analysis will provide the basis for either specific recommendations for water efficiency improvement or recommendations for further investigation.

The above presentation is intended as an introduction to the pulp and paper industry. In practice there is a wide variation in types of processes, products made, and equipment available, and no two mills are identical. Therefore it is wise to exercise caution when applying the above material to actual paper mill systems.

21

Arena: The food industry – wastewater auditing

21.1 INTRODUCTION

What is the nature of wastewater auditing? Recall the discussion of late in Chapter 3 where we stated that 'Closure can be sought for the whole audit domain (e.g. Figure 1.1), it can be sought between water inputs to the domain and the points of water usage. Also we might seek to close between the points of water usage and discharge from the domain…' The last sentence of the statement is a definition of wastewater auditing. In terms of Figure 3.2, a waste water audit could mean obtaining closure between the outputs of ' toilets and hand-basins ', ' Staff shower services ', ' Shop services ', ' air-conditioning ', the sewer discharge from ' wash-down ',' drinking fountain ' and ' discharge to the sewer '. In this example the discharges include wastes that are inimical to the environment and to public health, so the quality of the water is of great interest, as well as the quantities discharged. This is more generally the case with wastewater auditing. Often the quality and quantity of the discharges must meet certain criteria set by legislation or regulations. Thus the nature of a wastewater audit is much more akin to an environmental audit than is holistic water auditing. The desire to improve the quality of discharges leads naturally to the issue of wastewater treatment.

In this chapter we look at one small case study relating to soft drink manufacture. Then we consider some matters of importance in the food and meat packaging industries. We will not confine ourselves merely to wastewater auditing, but for the first time it will be a conscious focus in our discussion. Then we will continue with a summary of wastewater treatment processes, which are of interest in the packaged food industry generally. We think we are being fair by observing that the food processing and packaging industries have not had a good track record in their use of water and in their discharge of wastes.

We outline some reasons for engaging specifically in wastewater audits: more rigorous criteria for allowable discharges, the cost of land to install wastewater treatment facilities, the occurrence of overloaded treatment systems and the occurrence of odours of intensity outside of expectations. Added to these there are financial motivations too. We repeat the obvious cost of excessive use of water. There are the high costs of water treatment and the plant operating costs, the possibility of capital costs for future plant and the charges for disposal of waste in the sewer where available. Sometimes excessive product is lost in one or more water streams, imposing a direct increase in production costs. Taken together, these are powerful reasons for implementing a wastewater audit.

21.2 WATER AND WASTEWATER AUDIT PROCESS PRÉCIS

At the risk of appearing repetitive, we will highlight here major issues in the conduct of the audit, emphasising matters of importance to food processing industries.

21.2.1 Data Collection

The first major highlight is the sampling of wastewater streams for a range of qualities, which might include BCOD, SS, nutrient levels, heavy metal levels. Water quality assumes much more importance than in much other water auditing. Flow measurements retain great importance. Techniques for measuring flows in partially filled pipes and open channels assume a greater role. Variability is an issue in all water auditing. Yet in meat processing and packaging, for example, it might assume particular importance if processing is not continuous due to non-continuous supply of animals. We need to understand the process through which water passes as in all water auditing. The processes in the meat and packaging industries are not usually very complex. The procedures used for cleaning are at the heart of the production of wastewater streams. In some food processing industries solid materials make up part of the constituents of the final product. What are the procedures used for cleaning up spillages of these solids?

21.2.2 Simple water conservation measures

There is a small group of typical steps that can be taken in many meat processing and packaging enterprises to conserve the use of water. Sprays are typically used for hosing down floors or benches. A reduction of the nozzle size can often save significant amounts of water without impairing the function. Installing hand operated self releasing triggers can also save large quantities of water. Uncontaminated stormwater ought to be kept separate from contaminated waste streams, so as to limit the quantity of water entering wastewater treatment plant. Sometimes the stormwater can be gathered as a useful resource. Within the processing plant itself boilers and heaters or condensers almost always feature. The condensate stream can be treated and returned to the boiler. Opportunities should always be sought for using reject water from one unit operation in another needing lower quality water input. Savings achieved through such steps will both reduce the cost of water and the operating costs of existing wastewater treatment plant. If new wastewater treatment plant is installed, then the capital cost will be lessened.

21.2.3 Wastewater treatment

In some older works, the water auditor might find the situation where treatment of wastewater is necessary, and is not currently being carried out. In this situation the water auditor, in conjunction with a consultant, could devise both the processes to be used and the capacity of the wastewater treatment plant. Obviously the capacity will be selected with the savings identified above taken into account. Discharge from the plant will be to sewer, or sometimes to the environment, usually with the approval of an environmental regulatory agency and with strict conditions.

The starting point for the design of the wastewater treatment train is the specification of the quality of the water to be reused, discharged to sewer or to the environment. This quality will indicate appropriate treatments, as will factors like availability of land, capital, and time. A consultant will almost certainly be required. The processes will receive further treatment below.

21.3 CASE STUDY: A BOTTLED BEVERAGE WORKS

A significant bottled beverage works elected to undertake a water audit, though its track record of consciousness and practice of water conservation and was good. The basic goals of the audit were to quantify scheme water flows after their uses had been identified, to obtain closure at the conventional level of 10%, and to identify, and to evaluate financially, water conservation options. During the audit, following essentially the process outlined in our Chapter 3, water sources were identified as: scheme water from a water service provider, mineral water brought into the works, rainwater and bore-water. These were metered. The water sinks were soft drink, sewer, trade waste, gardens, storm water and evaporation. The soft drink going off site was readily measured, and sewer and trade waste lines were metered as were the gardens. Meters on the cooling tower and boilers enabled an estimate to be made of evaporation. A water flow diagram was constructed and it is presented in a modified form in Figure 21.1. Notice that in this case all trade waste was discharged to sewer without wastewater treatment.

Input flow quantities and output flow quantities were directly measured where meters were installed, calculated from these direct measurements where meters were not installed, or estimated based on already known specifications or parameters. Over a period of two weeks the flow meters were read daily, and averaged, resulting in mean flows expressed in litres per day.

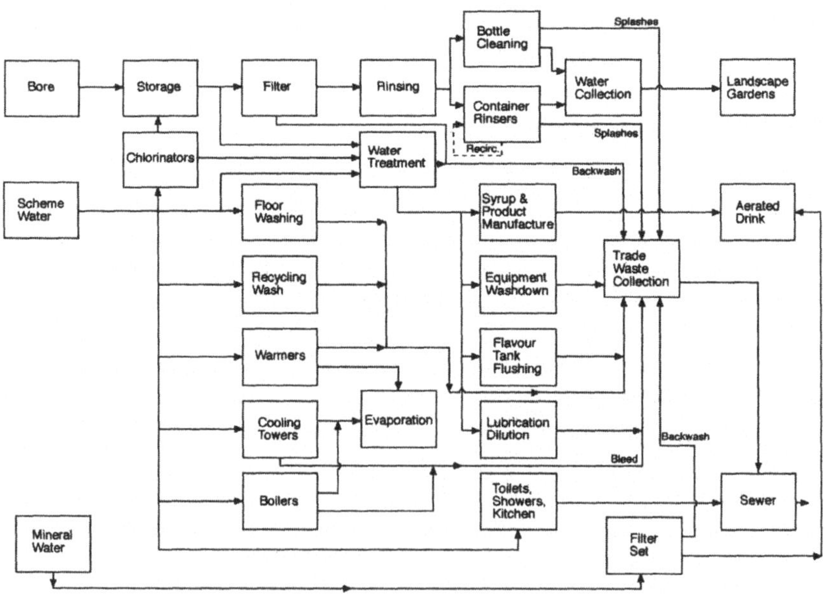

Figure 21.1 Typical water flow diagram for a bottled beverage works.

Daily readings of all the meters were made for a one-week period. The flows fluctuated significantly, implying that extending the monitoring period would have provided better data. We noted, though, that too long a period of monitoring was not possible, as the production of soft drink followed movement of the seasons. The flow measurements were used to obtain closure (see Figure 1.1); it was obtained to within 6%, a very satisfactory result. This gave us confidence to assess possible ways of saving water.

Our evaluation of options took account not merely of the economics of the options, but also of social and environmental factors, in accordance with the company's environmental management strategy. Water from the rinsing of containers provided the greatest opportunity for further water savings. At the time of the audit, this water was used for landscape irrigation. Diverting it to treatment for boiler water make-up and for cooling tower make-up water was the preferred strategy. As the amount of water recovered was far in excess of these needs, it was also applied to general wash-down uses and to toilet cisterns for flushing toilets. Readily available, shallow ground water was available to make up the deficit in water for the landscaping irrigation.

21.3.1 Benchmarks

Water audits are usefully repeated in time. The absolute values of the water used at various points in the water audit diagram are relatively meaningless unless scaled by either another appropriate quantity of other water or the quantity of the product. For example, the ratio of total scheme water use to product as a time series is a strong indicator of the relative water use efficiency of the plant. Comparison of the ratio of water use to product at the same time across a range of plants of similar sort provides some indication of the performance of this particular plant compared with others in the world. This is called bench-marking. Similarly, if we compare the ratio of water use of a particular unit operation with that of another unit operation, this can be a sensitive indicator of problems in a plant. The plant of the above case study was benchmarked against other similar plants in the world and found to be in the top quartile in terms of the ratio of the volume of scheme water use to product volume.

21.4 TECHNOLOGIES FOR TREATMENT OF WASTEWATER

21.4.1 Conventional Treatment Systems

Conventional treatment systems are a combination of physical and biological processes to remove biological oxygen demand (BOD) and suspended solids (SS) to the acceptable levels 9 (see Figure 21.2). This is achieved by removing the suspended solids by sedimentation and BOD by aerating the wastewater to satisfy the oxygen demand. The sludge is treated biologically and disposed of as a soil conditioner.

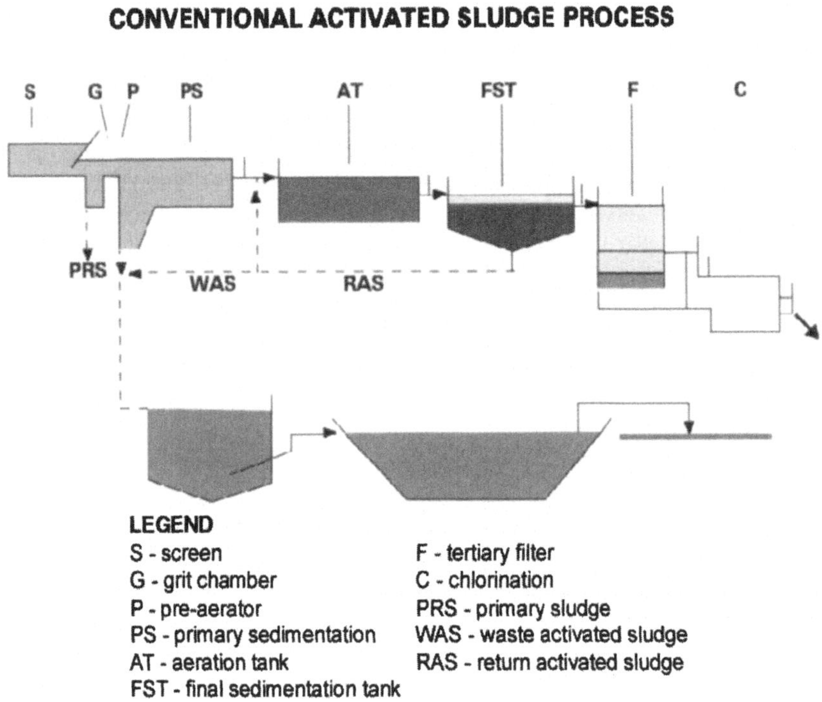

Figure 21.2 Schematic diagram of a conventional activated sludge wastewater treatment process (Adopted from UNEP 2002).

21.4.2 Primary Treatment

Primary treatments include preliminary treatment such as screening and grit removal and primary sedimentation. Flotation is sometimes included to remove grease and other floating matter. Chemical coagulation can also be incorporated to improve the efficiency of primary sedimentation which may be adopted during overloading. Pre-chlorination mainly to control odour is also considered as primary treatment.

Primary sedimentation is carried out in rectangular or circular tanks where the wastewater has a quiescent flow to allow the particulate matter to settle down. By this means 30-50 percent of the suspended solids accounting for about 30% of the total BOD will be removed.

The main design criteria for the sedimentation tanks are overflow rate, tank depth and detention time. Overflow rate is determined by the average daily flow divided by the surface area of the tank. This generally ranges between 16 to $32m^3/m^2$/day. The depth of the tank varies from 2 to 2.5 m. The detention time is more or less finalised by fixing the overflow rate and the depth of the tank so the detention time is between 1.5 to 3 hours, with an average of 2 hours. Shorter detention periods of 1/2 to 1 hour with less removal of suspended solids are also adopted depending on the capacity of the biological treatment systems such as the activated sludge process.

21.4.2 Secondary Treatment

The biological processes which are adopted for removing the organic matter which is in colloidal form or soluble. During this process the fine colloidal matter and dissolved organic matter are converted to settleable, flocculent matter. In most cases the biological process follows physical processes such as sedimentation, but some treatment systems, such as aerated lagoons and stabilisation ponds, can operate without the sedimentation process. Most commonly used biological processes are activated sludge, trickling filters, aerated lagoons and oxidation ponds.

21.4.2.1 Activated Sludge Process

The wastewater entering the aeration tank usually pass through the sedimentation process (see Figure 21.3). The wastewater flowing into the aeration tank contains organic matter both colloidal and dissolved, which is a good food supply for microorganisms. The flocs of bacteria consume the biodegradable organic substances in the presence of oxygen, producing new cell material and releasing carbon dioxide. The liquid in the tank is continuously directed to the secondary sedimentation tank where the heavy flocs of bacteria are separated. The clarified secondary effluent is designed to have BOD less than 20mg/L and suspended solids less than 30mg/L. The effluent is transferred for tertiary treatment or discharged to the environment as desired.

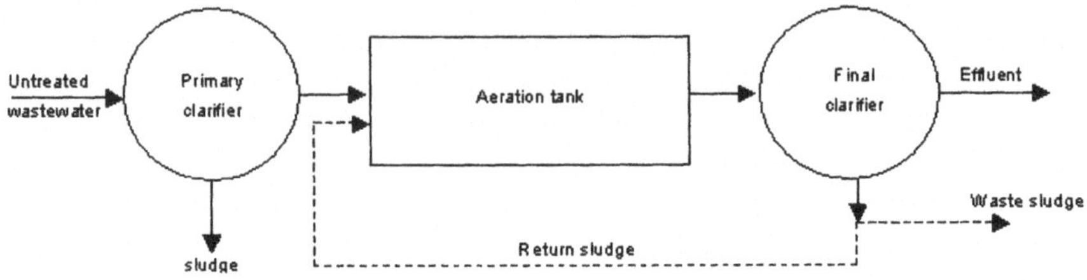

Figure 21.3 Typical schematic for a conventional activated process.

The settled sludge in the secondary sedimentation tank is recirculated continuously to the aeration tank to make sure that the mixed liquor suspended solids (MLSS) are hgh enough to treat all the incoming organic matter. The surplus sludge is directed for sludge treatment along with the primary sludge.

The aeration tank is aerated continuously. The rate of aeration depends on oxygen utilisation. BOD loading, aeration period and temperature are the main factors which determine the rate of aeration. This is generally between 10-30mg/L/hour to make sure that the dissolved oxygen level is not below 2mg/L.

Activated sludge treatment systems are highly mechanised systems. They are the most popular systems at present and suited for automated operations. The capital cost for building the plant, energy requirement for treatment and regular maintenance costs of operation are relatively high. The system also needs skilled technical personnel for technical support.

The activated sludge process can be operated in batches and is called a "Sequential Batch Reactor" (SBR). These systems are most suited for smaller flows such as most factories or institutions and not for municipal scale. The aeration tank is filled with wastewater and aerated to satisfy the oxygen demand. Then the activated sludge is allowed to settle and the water is allowed to flow out. Thus the tank is filled again to continue the operation. In most cases at least two tanks operating simultaneously is recommended.

21.4.2.2 Trickling Filters

A trickling filter is a bed of solid coarse media of stone, waste coal, gravel or different shapes of specially made plastic materials (see Figure 21.4). Wastewater is trickled over the media on which bacteria is attached on its surfaces. The process is not mechanical filtration or straining of the solids from wastewater. It is a process of removing the organic substances by use of bacterial action and hence called biological filtration. The filter medium is arranged in a tank on a support with openings to allow flow of air upwards by natural convection and treated wastewater to flow down to the tank.

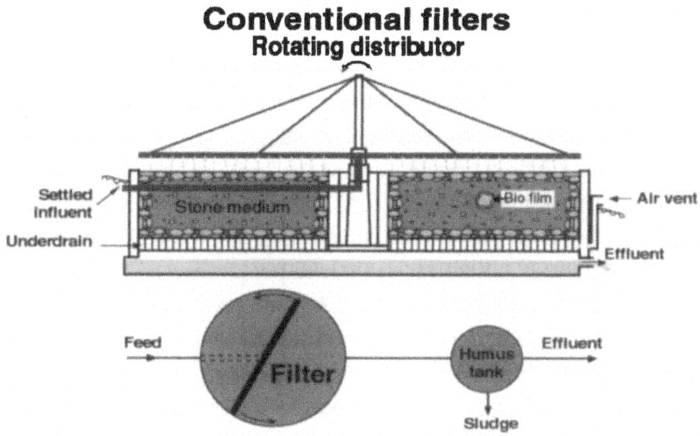

Figure 21.4 Schematic diagram of a trickling or biological filter (Adopted from UNEP 2002).

Trickling filter or Biological filters are not capable of treating the raw wastewater directly and the wastewater always receives primary treatment before being admitted to the system. The bacteria attached to the media in the form of bacterial slime, consume the organic substances. Bacterial slime receives the oxygen from the airflow through the bed of the media. The thickness of the bacterial slime increases as the bacterial grows and the heavy slime leaves the media along with wastewater. The solids in the bacterial slime are separated from the effluent by sedimentation and the sludge is directed to primary tanks for further treatment.

A two stage trickling filter, consisting of two filters and sedimentation units in series, is necessary to achieve the required effluent quality of 20mg/L BOD and 30mg/L suspended solids. Several options of recirculation are adopted in the systems to meet the requirements. The loading rate of the filter varies from 1-10 L/m^2/min so the area of the filter will be larger than the area of activated sludge. The depth of the media is 2-3m. The energy requirements and operating costs of a trickling filter is far less than activated sludge.

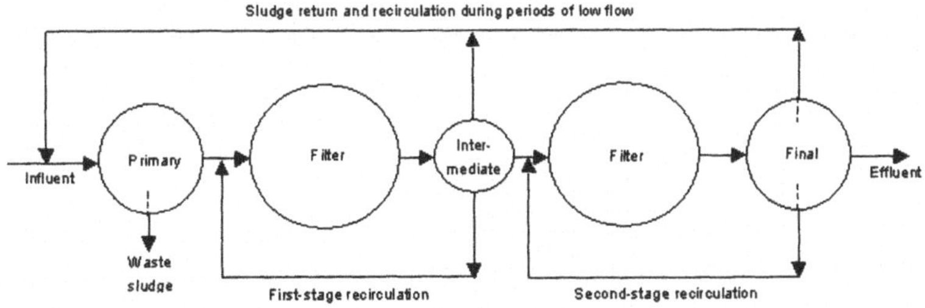

Figure 21.5 Typical flow diagram for a two-stage trickling filter plant.

21.4.2.3 Lagoons

Lagoons are very effective means of treatment of wastewater with respect to the removal of BOD, suspended solids and pathogenic organisms. They are more effective than activated sludge systems in the removal of bacteria, viruses, cysts of parasites and helminth eggs, due to sedimentation and die off because of long residence periods.

Lagoons are shallow basins of 1-3m deep. To avoid groundwater pollution, they are lined with plastic or clay for preventing percolation. They are cheap to construct and simple to operate, requiring about 10m^2 pond area per person (domestic sewage).

In the day time, algae, which may grow very well, produces lots of oxygen to keep the system aerobic (see Figure 21.6). In the night time, as there is no supply of oxygen, the bottom of the pond may become anaerobic. The ponds for which top is aerobic and bottom is anaerobic are called facultative ponds. In the pond, algae provide oxygen to the bacteria which consume organic matter and produce nutrients for algae. The loading rate to the pond is 0.02-0.05kg BOD/m^2/day with a retention time of 10-50 days. The treatment efficiency is increased with two or three lagoons in series then one single lagoon to treat the wastewater (Hammer 1997).

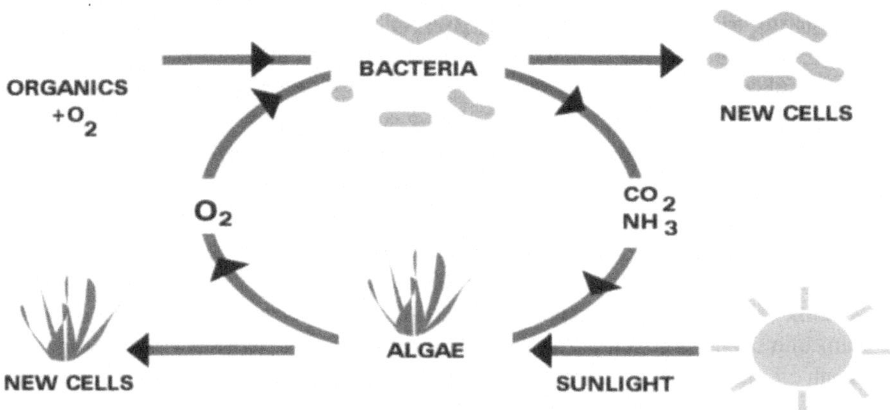

Figure 21.6 Symbiotic relationship between bacteria and algae in a wastewater lagoon (Adopted from UNEP 2002).

21.4.2.4 Sludge treatment and disposal

The source of sludge is the suspended matter in the wastewater and the biological mass generated during the treatment process. About 60% of the suspended matter is removed by primary sedimentation and the bacterial sludge produced is removed by secondary sedimentation tanks. Sludge is also generated at the on-site wastewater treatment systems in septic tanks.

The sludge will have a water content of 95-98.5 and hence dewatering is necessary for final disposal (Crites and Tchobanoglous 1998, Pierce *et al.* 1998). The sludge will have high BOD, nutrients and pathogens in the order of a million/100mL (Polprasert 1996). If industrial waste is also mixed with sewage, there will be heavy metals and toxic chemicals. As this will make the treatment and reuse of sludge difficult, it is not advisable to combine the sewage waste with industrial waste.

The treatment of sludge includes anaerobic digestion, sludge thickening, dewatering, drying, composting, reuse as soil conditioner. If the sludge is contaminated with heavy metals and other undesirable toxic pollutants, the disposal by controlled incineration or landfill in lined sites is adopted. Another method for disposal of contaminated sludge is a conversion process to produce oil from sludge (Bridle *et al.* 2000).

Anaerobic digestion is conducted in specially built digesters at 35°C with complete mixing, produces a mixture of methane and carbon dioxide. Methane is stored and used as an energy source. As the digested sludge has high water content of 94-99%, dewatering is essential for further treatment or disposal (Tebbutt 1998). The reuse of the water should be considered in the water conservation processes.

Composting is one of the processes which promote the reuse of sludge. Composting is an aerobic biological process for stabilisation of the organic matter under controlled conditions. The optimum conditions for composting are a moisture content of 40-60 percent, a carbon/nitrogen ration of 25 to 30 with pH of 6-8 and temperature 55°C. It is possible to adopt vermicompsting with a lower temperature by introducing earthworms to the sludge.

Final or ultimate disposal of sludge is to use the sludge as a soil conditioner, composted sludge as a fertiliser, production of oil from sludge (Bridle *et al.* 2000), land filling and incineration.

21.4.3 Tertiary Treatment

Tertiary treatment is defined as additional treatment needed to remove suspended and dissolved substances remaining after secondary treatment. The purpose of conventional treatment systems is to remove the organic matter in order to maintain a minimum dissolved oxygen level of the receiving waters. The substances remaining in secondary effluent are dissolved materials such as calcium, potassium, sulphate, nitrate, phosphate and synthetic organic compounds. Now the permit requirement of effluent discharge may include specific materials such as nutrients which cannot be removed by secondary treatment.

The main processes in tertiary treatment are processes to remove nutrients such as nitrogen and phosphorus, breakpoint chlorination for removal of pathogens, ion exchange to remove dissolved ions, and activated carbon filters to remove dissolved organic substances.

Phosphorus in wastewater originates from sanitary waste and detergents making a total of approximately 10mg/L in domestic wastewater. Maximum removal by primary and secondary treatment will be about 3mg/L resulting on average in 7 mg/L in the secondary effluent.

Phosphorus in wastewater can be removed by biological and chemical methods. As biological removal is complex, chemical precipitation is often employed. Alum is used, while increasingly being replaced by ferric ions (e.g. ferric chloride). Calcium ions are sometimes used to remove phosphorus too. These methods can reduce the phosphorus in wastewater to the required level of 0.01 to 1.0 mg/L.

Nitrogen can be removed by biological nitrification and denitrification. The concentration of nitrogen in wastewater will be about 40 mg/L N. A reduction of about 80- 95 % is possible by this process. As this is a biological process, reducing the concentrations below 4-5mg/L will be difficult.

Activated carbon is very effective in removing dissolved organics through adsorption. The Capacity of activated carbon can be determined experimentally by batch experiment. The removal efficiency of an activated carbon column can be calculated from the results of the batch experiment. Activated carbon is very efficient in removing chlorinated hydrocarbon and organic molecules which are difficult to remove by other means.

21.5 DISPOSING OF THE EFFLUENT

The food industry will have wastewater with very high BOD. It may be economical to treat the effluent partially before discharging to the sewer if the sewerage rate is paid as per the kg of BOD discharged to the sewer. If the final effluent is of BOD 250-300 mg/L it could be discharged to the sewer. If the BOD/suspended solid is 150 – 200 it is permissible for a deep water outfall discharge with the permission of the regulatory authority. If the BOD and suspended solid are 20 mg/l and 30 mg/l respectively the effluent can be discharged into any surface water course.

21.6 ECONOMIC SOCIAL AND ENVIRONMENTAL EVALUATION

It is the social and environmental responsibility of the industry to treat the wastewater to the required standard required by the regulatory authority. The extent of the treatment is also decided by economic factors. In many cases treatment and reuse on–site could be economical and environmentally most acceptable. This is a choice which depends on professional advice, scientific information and proper calculation and evaluation.

21.7 SUMMARY

The following summary table gives a general evaluation of wastewater treatment methods at a glance. The first portion is pertinent to the discussion in Chapter 12 also.

Table 21.1 Wastewater management technologies (Adopted from UNEP, 2002)

Technology	Capital cost	Operation & maintenance cost	Environmental impact
On-site technology			
Pit latrine	Low	Low	Pollution of groundwater
Composting toilet	Low	Low	Reuse of nutrients
Pour flush toilet	Low	Low	Pollution of groundwater
Improved on site treatment unit	Medium to high	Low to medium	Reuse of water and nutrients
Off-site technology			
Collection technology			
Conventional sewerage	High	High	Dependent on treatment
Simplified sewerage	Medium to high	Medium	Dependent on treatment
Settled sewerage	Medium	Low	Dependent on treatment
Treatment technology			
Activated sludge	High	High	Nutrients may need removal
Trickling filtration	Medium	Medium	Nutrients may need removal
Lagoons	Low to medium (dependent on cost of land)	Low	Nutrients may need removal; aquaculture can be incorporated
Land-based treatment	Low to medium (dependent on cost of land)	Low to medium	Reuse of water and nutrients
Constructed wetland	Low to medium (dependent on cost of land)	Low	Amenity value
Anaerobic treatment	Medium	Medium	Produces biogas; further aerobic treatment needed

21.8 REFERENCES

Bridle, T. R., Freeman, P., Luceks, A., Skrypsi-Mäntele, S., Ye, P. D. 2000. Production of high quality fuels using the enhanced Enersludge ™ process, *Wat Sci. Tech*, 41 (8): 45-51.

Crites, R. and Tchobanoglous, G. 1998. *Small and Decentralized Wastewater Management Systems*, WCB/McGraw-Hill, Singapore.

Hammer, M.J. 1977. Water and Wastewater Technology: SI Version, John Wiley & Sons, New York.

Peirce, J.J., Weiner, R.F. and Vesilind, P.A. 1998. *Environmental Pollution and Control*, 4th edition., Butterworth-Heinemann, Woburn, MA.

Polprasert, C. 1996. *Organic Waste Recycling: Technology and Management*, 2nd edition, John Wiley and Sons, Chichester.

Tebbutt, T.Y.Y. 1998. *Principles of Water Quality Control*, 5th edition, Butterworth-Heinemann, Oxford.

UNEP International Environmental Technology Centre 2002. *Environmentally Sound Technologies for Wastewater and Stormwater Management, An International Resource Book* – UNEP/ETC Technical Publications Services No. 15, IWA Publishing, London, UK.

22

Resources

The water auditor should consider stormwater as a resource. A section on stormwater (22.1) has therefore been included here and is based on one of the authors's contributions to UNEP (2002). The water auditor should also be aware of environmental water use and water resource management issues (22.3). This section also includes consideration of stormwater in urban areas. Finally the water auditor would benefit from knowledge of the technologies, energy requirements and cost of producing water from salty water (22.3). A glossary, specific to this chapter, can be found at the end of the chapter.

22.1 STORMWATER AS A RESOURCE

Stormwater in a community settlement is produced from house roofs, paved areas and from roads during rainfall events. In addition stormwater is produced from the catchment of a stream or river upstream of the community settlement. The amount of stormwater is therefore related to the amount of rainfall precipitation, and the nature of surfaces, with impervious surfaces producing more run-off. During a storm event the peakflow is higher and duration shorter with an impervious surface, while the peakflow is lower and duration longer with a vegetated surface (Figure 22.1).

Stormwater run-off may contain as much solids as household wastewater depending on the debris and pollutants in the path of the stormwater run-off, although in general the pollutant load of stormwater is lower than that of wastewater. Table 22.1 shows a comparison of urban stormwater sources and untreated sewage in North America.

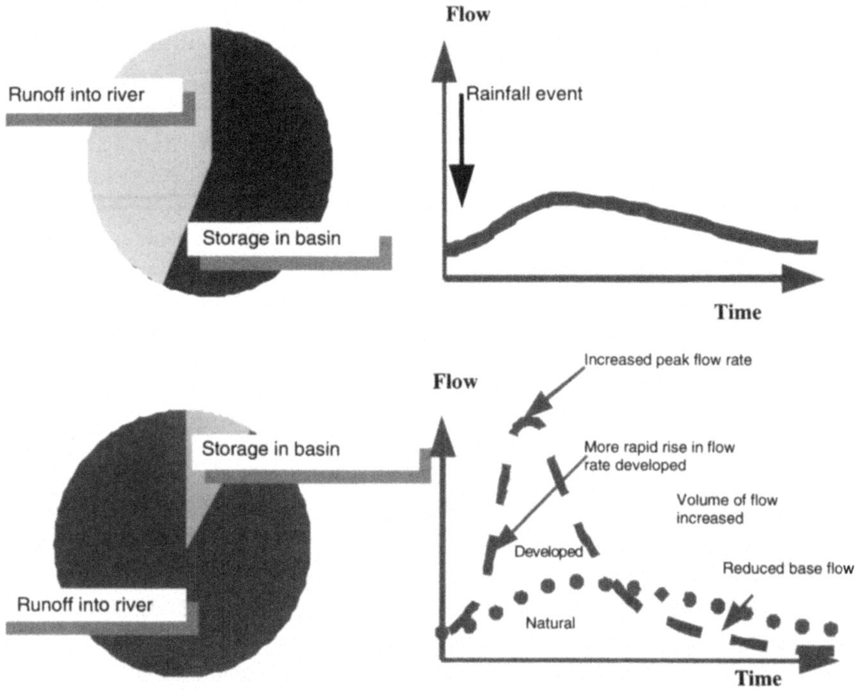

Figure 22.1: Rainfall runoff relationship showing two different surfaces (impervious and natural). (Adopted from UNEP, 2002).

22.1.2.1 Impact of stormwater

Solids in stormwater form sediments and can eventually clog drains, streams and rivers. Grease particles form scum and are aesthetically undesirable. The nutrients N and P cause eutrophication of water bodies, with lakes and slow moving waters affected to a greater degree than faster flowing waters. In the former the algae which are fertilised by the nutrients, settle as sediment when they decay. The sediment acts as a store of nutrients and regularly releases the nutrients to the water column, thus the cycle of bloom and decay of the algae is intensified. In the early stages of eutrophication aquatic life is made more abundant, because fish, for example, graze on the algae. With too high a concentration of algae, the decaying algae contribute to

BOD and the water is deoxygenated. Thus wastewater, which has been treated to reduce BOD but still high in nutrients, can still have a significant impact on the receiving water. Some algae produce toxins which can be harmful to bird life and irritate skins coming into contact with the water. Eutrophic water adds to the cost of water treatment, when the water is used for drinking purposes.

Table 22.1 Comparison of the characteristics of stormwater sources and untreated sewage (Novotny and Olem, 1994; Novotny, 1995)

Type of wastewater	BOD_5 (mg/L)	Suspended solids (mg/L)	Total N (mg/L)	Total P (mg/L)	Total Coliforms (MPN/100mL)
Urban stormwater	10-250 (30)	3-11,000 (650)	3-10	0.2-1.7 (0.6)	10^3-10^8
Construction site run-off	NA	10,000-40,000	NA	NA	NA
Combined sewer overflows	60-200	100-1,100	3-24	1-11	10^5-10^7
Light industrial area	8-12	45-375	0.2-1.1	NA	10
Roof run-off	3-8	12-216	0.5-4	NA	10^2
Untreated sewage	(160)	(235)	(35)	(10)	10^7-10^9
Wastewater treatment plant effluent (secondary treatment)	(20)	(20)	(30)	(10)	10^4-10^6

Figures in brackets = mean values; NA = not available; MPN = most probable number

Other pollutants in stormwater are heavy metals and possible toxic and chemical hazardous substances. Heavy metals include copper, zinc, cadmium, nickel, chromium and lead. The content and concentration are dependent on the type of materials used for roofing and guttering. In high enough concentrations these heavy metals are toxic to bacteria, plants and animals, and to people.

Spills of chemicals, particulates from motor vehicle exhaust and deposition of atmospheric pollutants can contaminate stormwater. These pollutants will affect downstream receiving waters, and treatment systems if the stormwater is treated.

Contaminated stormwater can contaminate groundwater. This is through infiltration of the stormwater through the soil to unconfined groundwater aquifer. Soil can filter some pollutants, but soluble pollutants (e.g. nutrients and heavy metals) and very small particles (e.g. virus) travel with the water to the groundwater aquifer.

Heavy storm events can cause flooding. The effects of flooding can be severe. Water levels in drain, stream and rivers rise considerably and the flow of water can erode soils and embankments. Sediments which have been deposited in quiescent stretches of a stream can be resuspended and transported further downstream. In urban areas the water picks up litter and solid wastes in its path as well as other diffuse pollution sources, and spread these in the downstream flooded areas. Aquatic environments and water-fowl habitats can be destroyed, and these may take some time to recover. The amenity value of these, as well as recreational lakes, is therefore degraded. Engineered structures, such as culvert and bridges, can be choked with wastes and debris, causing more wide-spread flooded areas.

22.1.2.2 Natural purification processes

Before considering technologies for stormwater management it is instructive for us to examine natural processes that cycle waste materials. In nature waste materials are produced by living organisms (plants, animals and people). These wastes include faecal materials, leaf litter, food wastes and dead biomass. Yet streams and rivers flowing through a pristine forest, or freshwater lakes in a forest, have generally an excellent water quality. Thus there are natural processes, which purify the naturally produced wastes. These wastes are characterised by their organic nature (that is derived from living or once living organisms). They consist of carbon, nitrogen, phosphorus and other elements, which constitute the building blocks of living organisms. These elements are continuously cycled in nature. Three of them (carbon, nitrogen and phosphorus cycles) and the water cycle are relevant to stormwater management. Figure 22.2 shows the natural carbon cycle.

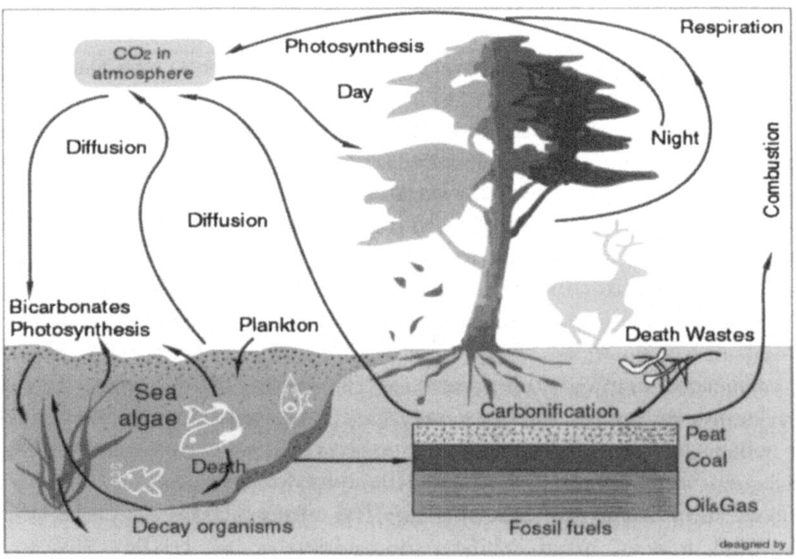

Figure 22.2 Carbon cycle. (Adopted from UNEP, 2002).

The following transformation processes occur in the carbon cycle. Plants photosynthesise glucose from carbon dioxide gas and water, and in turn more complex organic matter is synthesised. Plants are consumed by plant-eating animals, which in turn are consumed by meat-eating animals. Organic carbon compounds are digested by these animals and re-synthesised into other forms, which are useful for energy, cell growth and cell multiplication. Carbon dioxide is released into the atmosphere during the process of respiration. The respiration process releases energy for the organism through oxidising the organic carbon. Plants and animals produce waste materials and will eventually die. Leaf litter, animal wastes and dead organic matter are decomposed by bacteria and other decomposers releasing the carbon as carbon dioxide thus completing the carbon cycle. Oxygen is required in the process of respiration and oxidation of organic carbon, and this is the reason for the oxygen demand of organic wastes. Some organic matter from dead animals and plants is, however, stored in nature, particularly in sediments, and slowly turns into peat or more stable carbon-rich materials.

In the process of decomposition not only is carbon released as carbon dioxide, but other minerals are released. These minerals are involved in other cycles, such as the nitrogen cycle (Figure 22.3) and phosphorus cycle (Figure 22.4).

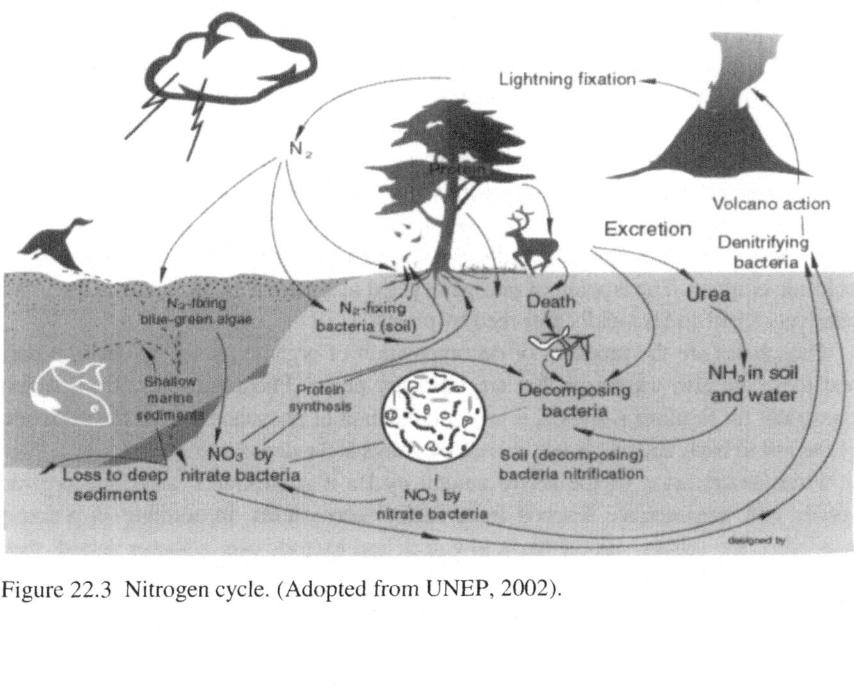

Figure 22.3 Nitrogen cycle. (Adopted from UNEP, 2002).

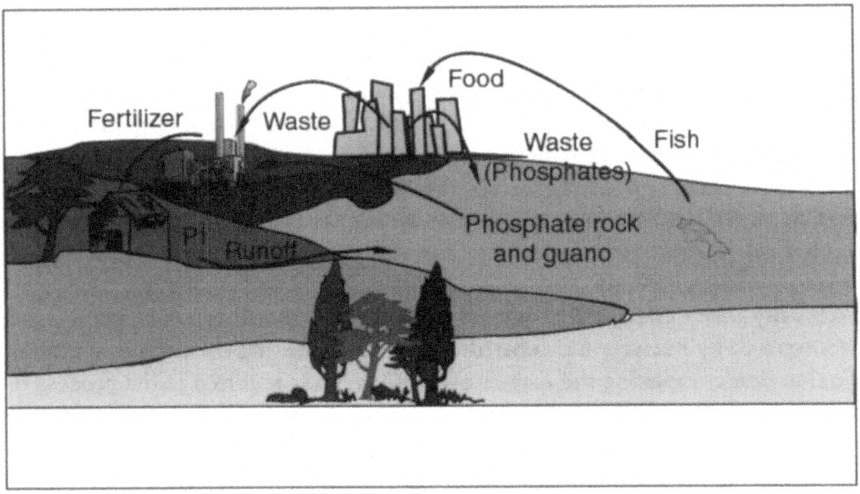

Figure 22.4 Phosphorus cycle. (Adopted from UNEP, 2002).

Ammonia is generally the form of nitrogen released from the decomposition of organic wastes. Provided that oxygen is available the ammonia is oxidised by a group of bacteria (termed nitrifiers) to nitrate. This process is another that exerts oxygen demand on the environment. Nitrate is the form of nitrogen that is normally taken up by plants for protein synthesis. Nitrate may on the other hand, under conditions devoid of oxygen (anaerobic conditions), be converted by a group of bacteria (termed denitrifiers) to nitrogen gas. Denitrification generally takes place in sediments, where anaerobic conditions and availability of organic carbon promote the process.

Nitrogen gas in the atmosphere is very large in quantity, but is inert. Relatively small quantities are converted into forms that can be utilised by plants. These are converted through the activity of nitrogen-fixing bacteria in the root-nodules of some plants, nitrogen-fixing blue-green algae or through lightning. Some is contributed by volcanic eruption. The amount of nitrogen cycled in a natural environment is therefore relatively small and is rapidly absorbed by plants.

Phosphates are the products of decomposition of organic matter by decomposers and these are also the forms that are taken by plants. Phosphate rock, from which phosphate for fertiliser is mined, is an accumulation of phosphorus from the excretion of the guano birds and that is not utilised by plants at the deposition site.

From examination of the above natural cycles it is clear that very little organic wastes and nutrients are leached from natural ecosystems. In addition in a forest ecosystem the surface run-off has a low peak and extends over a longer period, thus solids are filtered from the water, and nutrients have a higher likelihood of being absorbed by plants. The soil in a forest ecosystem can provide additional purification processes. Soil bacteria will consume organic carbon and reduce BOD. Soil minerals (particularly clay minerals) can adsorb metals and phosphates. Plant roots take up nutrients released by bacterial decomposition from water percolating through the soil.

Pathogens, if any, generally die-off, because of unfavourable conditions outside their hosts for an extended period and competition with naturally occurring mirco-organisms. The water cycle therefore produces surface water and groundwater of very high quality (Figure 22.6).

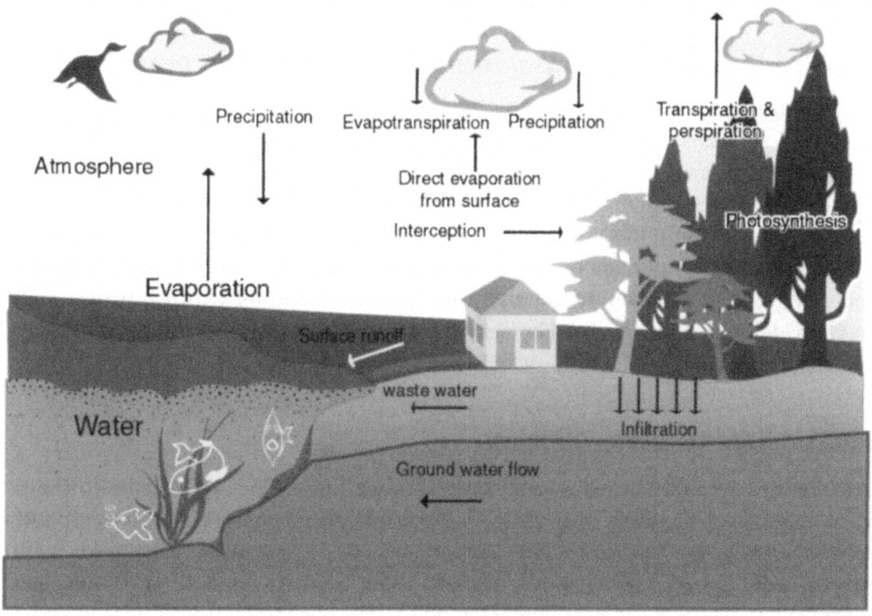

Figure 22.6 Water cycle. (Adopted from UNEP, 2002).

The natural cycles (also termed biogeochemical cycles) can provide an insight into the natural basis of stormwater management. For disposal of stormwater into a natural ecosystem, as long as the natural purification capacity of the ecosystem is not exceeded, we can rely on the existing natural processes to assimilate the wastes without degrading the quality of the environment. On the other hand once the natural capacity is exceeded, engineered systems are required. There is no reason, however, why the same physical, chemical and biological processes taking place in nature cannot be used as a basis for technology development and for waste management.

We note that in nature the cycling of the elements provide a pathway for reuse of the materials in the wastes. We should consider how we can use the same processes to recycle stormwater. A limitation of natural purification processes is that they can only handle naturally occurring wastes. The latter can include human wastes, but not toxic chemicals that stop the natural processes. In addition a large human settlement removes a large area of natural ecosystem and generates a large amount of wastes, and the combination of the two rapidly and significantly impact on our natural environment. Clearing of vegetation reduces evapotranspiration, while roads and houses introduce impervious surfaces. Consequently rainfall run-off has a higher peak and is generated rapidly, promoting local flooding (Figure 22.1).

22.1.2.3 Sustainable versus unsustainable stormwater management

The natural purification processes and biogeochemical cycles described above provide a basis for determining what is environmentally sustainable management practices for stormwater. Discharge of stormwater into an environment exceeding the natural purification capacity of that environment will result in the accumulation of organic materials (carbon), nitrogen, phosphorus or other pollutants that cannot be absorbed by the ecosystem constituting the receiving environment. Accumulation of organic materials will result in a high oxygen demand that cannot be met by oxygen transfer from the atmosphere. Undesirable anaerobic conditions are a consequence.

22.1.3 Stormwater collection

As mentioned earlier, stormwater flows through the landscape's natural drains. Piped stormwater collection was a development in European cities to overcome odour and improve aesthetic appearance of wastewater disposed with stormwater. The covering of ditches used for combined sewerage and stormwater was an intermediate step in using natural drainage to construct sewerage for combined wastewater and stormwater. With separate collection of wastewater there is an opportunity to return some stormwater flow paths to more natural states to improve urban amenity value.

22.1.4 Stormwater treatment

Treatment of stormwater means the removal of pollutants from the water. The first principle to bear in mind therefore is to prevent pollutants from entering the water in the first place. We need to ensure that surfaces through which stormwater run-off passes over should as far as possible be free from solids and other wastes. Thus collection of solid wastes is an important part of stormwater treatment or its prevention. Separate collecting of wastewater and stormwater also belongs to this principle.

Besides preventing pollutants entering the water, water conservation means that less volume of water has to be treated. Since the size of treatment systems is primarily governed by the volume of water to be treated rather than the amount of pollutants in the water, less volume means smaller treatment plants and corresponding capital cost.

Stormwater can be polluted as discussed above. When collected in a combined sewerage system it is treated with the wastewater, though treatment is not effective during peak heavy stormwater run-off periods resulting in combined sewer overflow (CSO) that is not treated. Storage basins or tanks can be used to accommodate moderate peak flows of combined stormwater and wastewater, and treating the stored water at night when wastewater flow is a minimum.

Separately collected stormwater is generally treated by passing it through a settling basin to remove solids (Figure 22.7). The retention time in the settling basin is designed so that solids can settle in say 20 minutes for a one in five year storm-event. For storm-events less than the design value removal efficiency is greater, while for storm-events greater than the design value removal efficiency is lower. Mechanical devices have been developed that can trap gross solids. Both settling basins and mechanical traps need to be cleaned regularly to maintain solids removal efficiency.

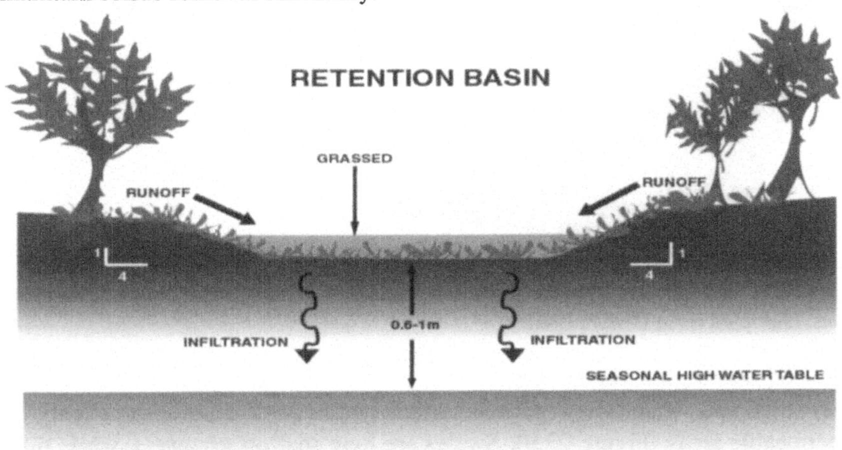

Figure 22.7 Stormwater treatment by settling. (Adopted from UNEP, 2002).

Naturally landscaped stormwater drains can help filter out fine sediments through the action of vegetation slowing down the flow and trapping solids. Permeable surfaces allow rainwater to percolate into the soil, thus treating the water in much the same manner as land based treatment of wastewater and at the same time reduce the amount of run-off. Pavements have been designed and manufactured for this purpose. Directing run-off to vegetated area (rainwater harvesting) can reduce down-stream flow and reuse the water for maintaining plant growth. This is especially beneficial in arid climates. Four techniques for stormwater treatment are described below. Used judiciously these can treat stormwater locally (at source, Figure 22.8). Applying these on a sub-catchment scale (site), or whole catchment scale (region) can reduce flooding and undesirable impacts of stormwater, while at the same time improve the amenity value of the landscape through creation of, for example, passive recreation water bodies.

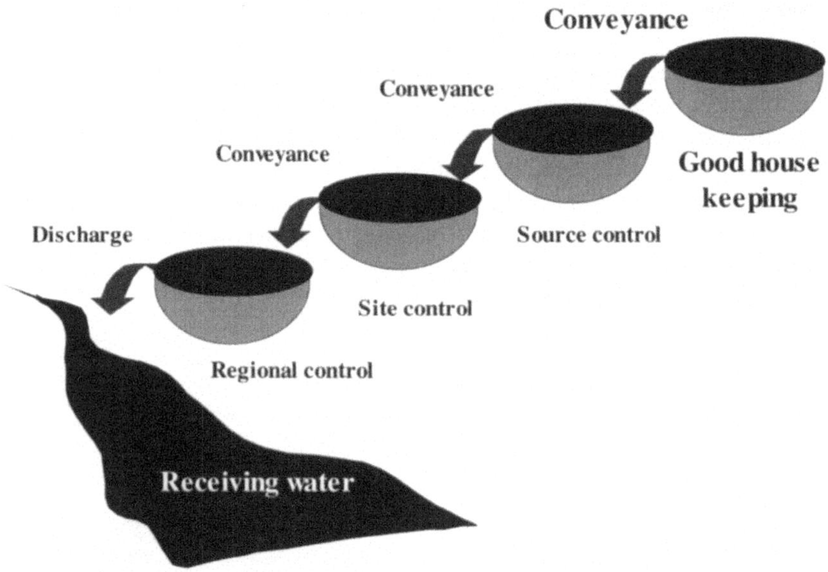

Figure 22.8 Management train for stormwater at the local, sub-catchment and catchment levels. (Adopted from UNEP, 2002).

22.1.4.1 Filter strips and swales

Filter strips and swales are vegetated surface features that drain water evenly off impermeable areas (Figure 22.9). Swales are long shallow channels, while filter strips are gently sloping areas of ground. They allow run-off to flow in sheets through vegetation, slowing and filtering the flow. Swales also act to temporarily store and infiltrate the run-off into the ground. Sediments are removed from the water, and vegetation can take up any nutrients in the water. Swales and filter strips can be integrated into the surrounding land use, for example road verges. Local grasses and flower species can be introduced for visual effect and to provide a wildlife habitat. Maintenance consists of regular mowing, clearing litter and periodic removal of excess silt.

Figure 22.9 Filter strip and swale in an urban landscape. (Adopted from UNEP, 2002).

22.1.4.2 Filter drains and permeable surfaces

Filter drains consist of permeable materials located below ground to store run-off. Run-off flows to the storage area via a permeable surface (Figure 22.10). The permeable surface can be in the form of grassed or gravelled areas, paving blocks with gaps between individual units or paving blocks with vertical voids built in. Water is therefore collected from a large surface area, stored in the filter drains and allowed to infiltrate through the soil. The permeable fill traps sediments and thereby clean the run-off. Filter drains and permeable surfaces are currently used for road verges and car parks. The surfaces should be kept clear of silt and cleaned regularly to keep the voids clear. Weed control may be necessary.

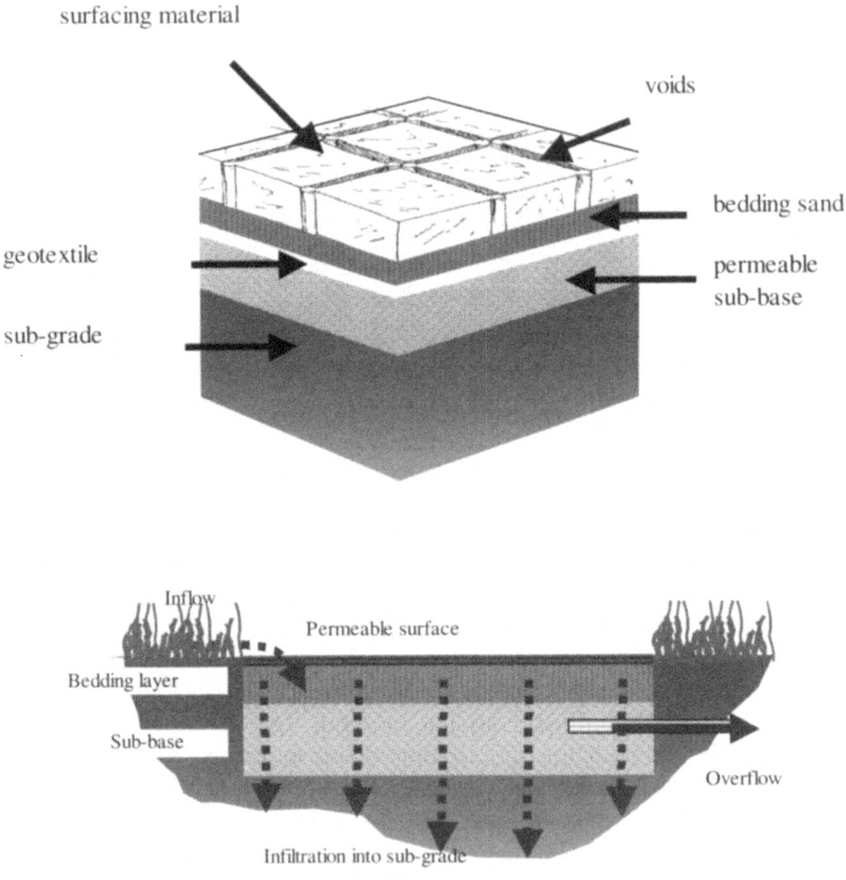

Figure 22.10 Permeable pavements. (Adopted from UNEP, 2002).

22.1.4.3 Infiltration devices

Infiltration devices drain water directly into the ground. They include soakways and infiltration trenches, which are located below ground, and into which stormwater run-off is directed. They function by storing water and allowing the water to infiltrate into the ground. Figure 22.11 shows a cross-section through a traditional soakway or a chamber soakway. They work well when the soil is permeable and the groundwater table is not close to the surface. Maintenance consists of regular inspection to ensure the infiltration capacity is maintained. Areas draining to an infiltration device should be kept clear of silt, as this will get washed into the device and reduce its permeability as well as filling up space that should be used for storage.

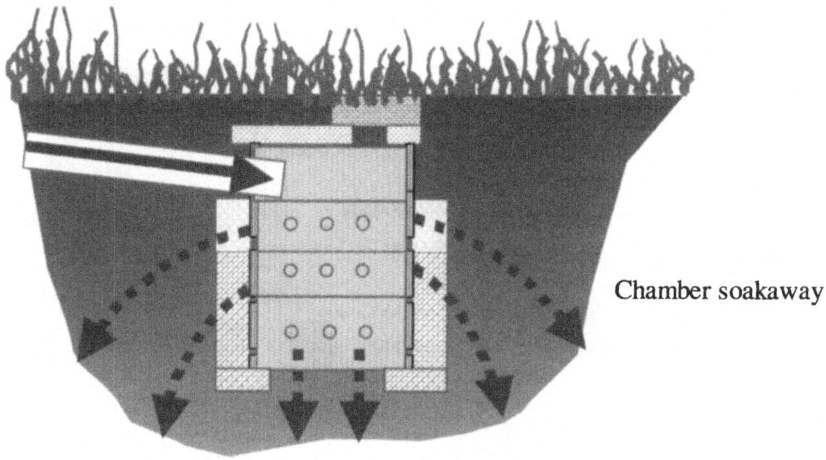

Figure 22.11 Cross-section through a traditional soakway or a chamber soakway (CIRIA, 1999)

22.1.4.4 Basins and ponds

Basins are areas for storage of run-off that are dry during dry weather, whereas ponds have permanent water (Figure 22.12). Both act to store water and therefore attenuate the flow of water during a storm. Flow downstream of the basins or ponds can therefore be controlled. Basins and ponds also act as infiltration devices. Basins and ponds are usually used at the end of a train of treatment for stormwater, and provide additional step if source control does not have an adequate capacity to control run-off. Detention time is of the order of two to three weeks. Both basins and ponds can be vegetated, so that we can have a range of features, including wetlands that have amenity values for passive recreation or wildlife habitat. Run-off water quality is improved upon storage in basins or ponds because of sedimentation of solids, bacterial action and nutrient uptake by vegetation. Water stored in ponds can also be used for irrigation of parks and gardens or for fire-fighting and other purposes. Basins and ponds need to be maintained to control vegetation and removal of accumulated silt.

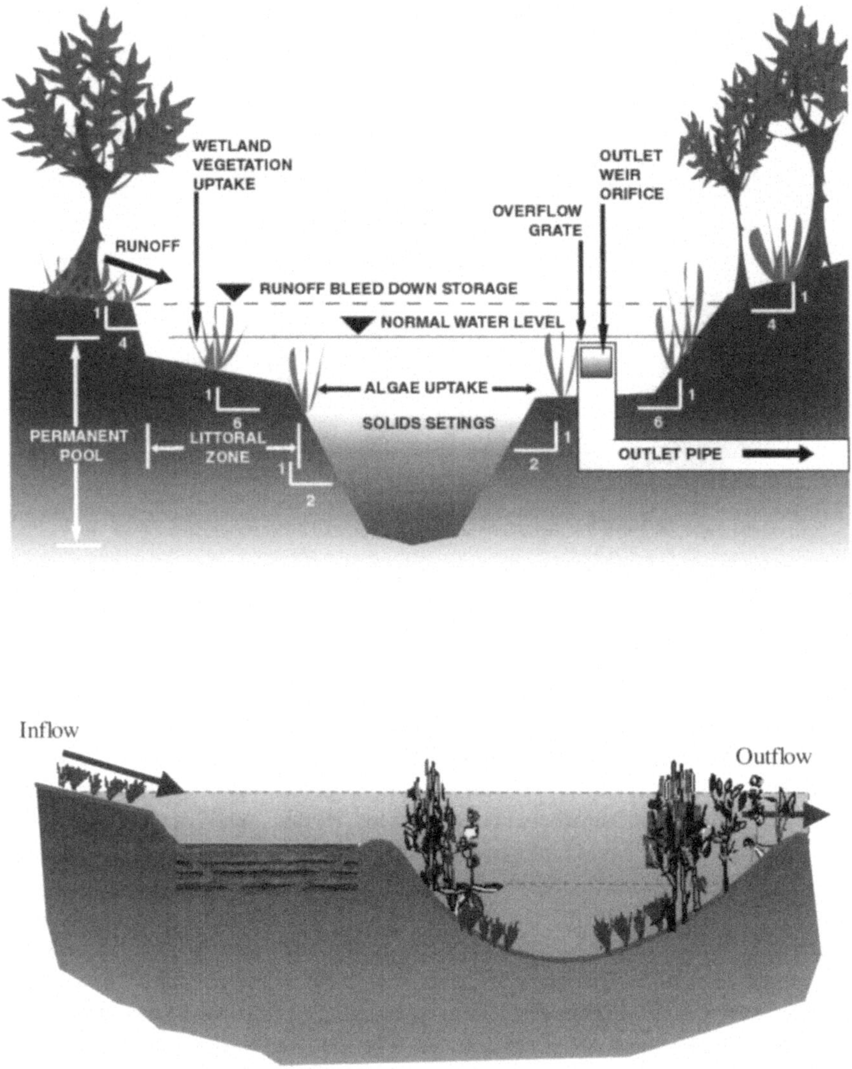

Figure 22.12. Pond, basin and constructed wetland for stormwater treatment. (Adopted from UNEP, 2002).

In Table 22.1 below the appropriateness of several technologies for different arenas and their costs are outlined.

Table 22.1. Technologies for stormwater management

Technology	Source control	Site control	Regional control
Filter strips and swales	√		
Filter drains and permeable surfaces	√		
Infiltration devices		√	
Basins and ponds			√

*Cost increases from source control to regional control technology.

22.1.5. Stormwater use

Stormwater is generally of a higher water quality than wastewater. Use of stormwater can take place at household, municipal or larger (regional) scales if desired. Use at the household and municipal levels is described below.

22.1.5.1 Household level

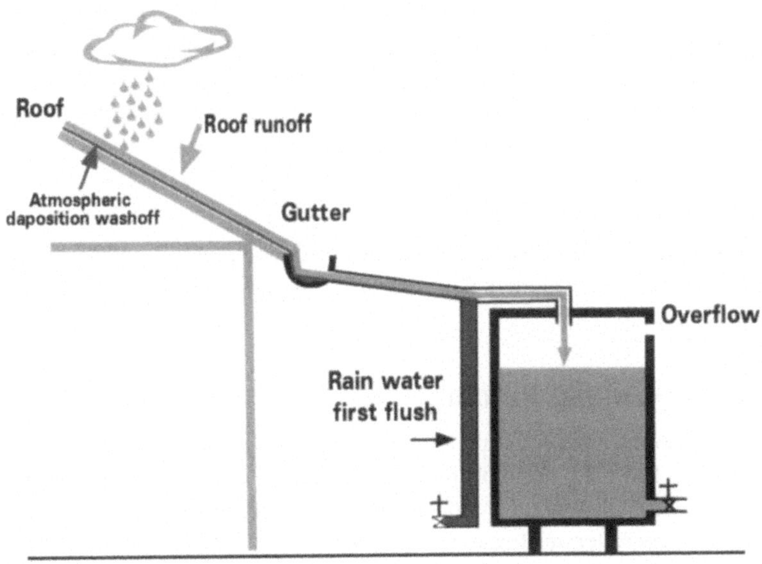

Figure 22.12. Diverter for the first flush from roof run-off. (Adopted from UNEP, 2002).

Householders can use stormwater by collecting roof run-off in a tank for use as drinking water (common in arid regions), flushing toilets or for irrigation of the garden. The first flush of roof run-off can be contaminated by dust particles, leaf litter and animal droppings. The first flush can be simply diverted using a simple diverter (Figure 22.12). A screen can be placed at the inlet to the tank to filter gross particles. Water for drinking will still need to be boiled to denature pathogens.

Water from the roof can be directed to the garden beds directly rather than through soakways, and in this way shallow rooted vegetation can benefit from the water, especially in arid regions.

22.1.5.2 Municipal level

At the municipal level stormwater can be stored in ponds for use for irrigation of parks and gardens and for fire-fighting purposes. This is in addition to employing the ponds for flood control and for improving the amenity value of the water. Other uses are for groundwater recharge, either as a means of storing water, e.g. during the rainy season, for withdrawal in the dry season. Groundwater recharge can also be used to prevent encroachment of sea water near the coast where there is heavy groundwater withdrawal in excess of natural replenishment by precipitation.

Stormwater collected from community buildings (i.e. government and churches) may be stored in reservoirs to be used by the community during dry periods. This method is used in the Pacific SIDS atoll countries of Tuvalu and Kiribati.

22.1.5.3 Water auditor's role

The water auditor will promote the use of stormwater whenever possible considering the environmental, economic, social, cultural and institutional setting of the community.

22.1.6 Stormwater disposal

Disposal of stormwater should preferably be considered only when reuse options are not feasible. Ultimate disposal for stormwater is onto land (by infiltration to groundwater) and to water environments (river, lake, ocean). These have been covered as part of stormwater treatment (22.1.4) and reuse (22.1.5), because they utilise infiltration as a general technique. Techniques for reuse are those that delay its ultimate flow to water environments to improve flow management and hence reduce the frequency and extent of flooding. At the same time these techniques also generally remove pollutants (particulates and oils) prior to the water reaching a river, lake or the sea, while creating amenities such as wetlands, waterfowl habitats and water-based passive and active recreational facilities.

22.1.7 Sound Practices

22.1.7.1 Overview of Sound Technologies and Practices

Technologies which are environmentally sound are technologies which help protect the quality of the environment. It may be argued that technologies used to manage stormwater are inherently environmental technologies, because without these technologies the pollutants in stormwater will negatively affect the environment. Some of these technologies may utilise less energy than others, produce less air pollution or hazardous products. Hence some of these technologies are more sustainable. The application of a technology is dependent on local physical factors of land availability, its topography, climate, soil, availability of energy and existing land uses. Sound technology practice is therefore dependent on being able to fit the technology to the local conditions.

Sound practice is also dependent on the context of the local community where the technology is to be applied. Long term sustainability is a function of community resources (funds, skills) to afford the technology and its willingness to pay for the technology and its operation. Sound practices are therefore practices which fit into the environmental, economic, social, cultural and institutional setting of the community.

22.1.7.2 Environmental considerations

Achievement of protection of environmental quality is implicitly assumed when we consider technologies for stormwater management. These considerations are (i) the need to protect the environment and (ii) the imperative of recycling/reusing the water and any nutrients in the water. Each local environment has its own capacity depending amongst others on the natural throughflow of water, climatic, vegetation and soil conditions.

Reuse of the water and nutrients conserve these resources in a world where water will in the future be a precious resource for growing food and maintain ecosystems for the world's increasing population and standard of living. Reuse of water can in fact fulfil the objective of protecting the environment, because reuse has standards which have to be met prior to the water being able to be reused. Treatment, reuse or disposal of stormwater containing toxic and hazardous chemicals will be considerably more difficult than treating the toxic and hazardous wastes separately.

22.2 ENVIRONMENTAL WATER USE AND WATER RESOURCES MANAGEMENT

22.2.1 Introduction

Pressure on our water resources is increasing due to a range of factors, in particular rising population levels. Global water use, by humans, tripled between 1950 and 1995 (Anderson, 1995). It is expected that within 25 years at least 65 nations will be seriously short of water (Fullerton, 2001).

This pressure is increasing competition between demands, for example between industry, irrigated agriculture and the urban populations. At the same time, growing awareness of the needs of the environment is pushing governments to regulate to ensure environmental requirements are met, at least in part.

Cities increasingly have to compete with the requirements of both the environment and irrigators for water supplies. There is growing awareness that the current levels of extraction from many catchments and rivers are unsustainable (Commonwealth of Australia, 2002).

Our knowledge of the environmental requirements for water remains inadequate. However, it is clear that in many areas of the world we are using far more water than we should. By the mid-1990s, for example, Libya was using, annually, 374 per cent of its available water supply. In short, its groundwater sources were being mined (Anderson, 1995).

In addition to the increasing water requirements in terms of quantity, water quality is coming under pressure. Human activities adversely impact on water quality which can significantly affect the environment, as well as rendering water sources unfit for further use.

While traditionally water has been managed principally to protect health and safety and to prevent damage and danger to people and property, it is now recognised that water should be treated as a valuable resource. In the urban environment this position is particularly relevant for stormwater, which is increasingly being considered as an asset. There is also the increased value being placed on the recreational and environmental values of pristine catchments and waterways.

These factors point to the need to consider water use and water management in the context of sustainability. In order to achieve this, we need to be able to determine the sustainable yield of a water resource and to ensure that abstraction does not exceed this yield. However, sustainable water use will vary from site to site, and also from year to year in response to other factors such as climate. Furthermore, sustainable water use needs to be viewed within the context of all social, environmental and economic considerations.

This section explores the issue of environmental water use and water resource management in the context of stormwater and the urban environment. The role of water auditing will not be defined explicitly in this section. However, issues relating to environmental water use and water resource management where auditing has a role will be highlighted.

22.2.2 A balancing act

Given the increasing pressure on water resources and increasing water use, people are starting to look to the future to a time when, potentially, fresh water will no longer be as readily available.

Of all water on earth, only about 2.5 per cent is fresh and most of this (76 per cent) is located in glaciers and ice caps, with 23.5 per cent in groundwater. Only 0.5 per cent of global water is in freshwater lakes, rivers, water vapour in the atmosphere, and water stored in animal and plant life, and in the soil (Smith, 1998). This tiny portion of global water is what humans have available to use, and more than half of earth's accessible fresh water is now being used, either directly or indirectly (Lowe, 2002).

The concept of sustainability or sustainable water use has now been used by policy makers and scientists for nearly two decades. "Sustainable development" became the catch phrase towards the end of the 1980s (Brundtland, 1987), the idea being that we use our resources in such a way that will meet the needs of both the current and future generations.

The reality is that sustainability is something of an abstract concept as we are not meeting the needs of the current generation in many parts of the world and it is difficult to determine what the needs of future generations will be. Furthermore, we do not know with certainty what a sustainable yield is, particularly given the severely modified environment in which we now live.

For example, many riverine environments have been modified through diversions, dams, weirs and extraction. The Murray River in Australia is currently flowing at 27 per cent of its natural capacity with consequential severe environmental impacts, including the closure of its mouth. The capital city of the State of South Australia, Adelaide, relies on the Murray River for much of its drinking water and it has been estimated that, unless river flow is restored to something like at least two-thirds of its natural flow rate, by 2020 the water will be too saline for human consumption two days out of five.

Therefore, it is important to consider the environmental requirements when determining a sustainable abstraction rate. However, this consideration is something of a balancing act.

While under ideal conditions ecological requirements should be considered and addressed prior to allocating water for economic or social needs, the reality is that in many parts of the world this has not occurred. There will always be times when economic or social issues will take precedence over the environment, often for political reasons. The main objective is to find an appropriate balance between these competing needs. Management of groundwater levels in Perth, Western Australia, provides a good case study to demonstrate this balancing act.

22.2.2.1 Case Study – Groundwater levels in Perth, Western Australia

Western Australia has been experiencing over 25 years of generally below-average rainfall, with the years 2001 and 2002 particularly dry. This has considerably reduced reservoir inflows, lowering water levels in the dams that constitute a large proportion of the scheme water supply to Perth, the capital city, and increased public abstraction from the city's groundwater sources, to the extent that by 2003 over half of the scheme water was coming from groundwater.

However, recharge rates to the aquifers have also been reduced with the drought conditions and groundwater levels have declined, with consequent adverse effects on a range of ecosystems such as wetlands.

Environmental conditions have been set by the Minister for the Environment through the Environmental Protection Authority to ensure certain minimum groundwater levels are maintained. Close to 20 of these conditions have been breached and a number of public water supply bores have been shut down to ease localised pressure on the groundwater levels near environmentally sensitive areas. In addition, at the beginning of the 2001/02 summer, water restrictions were instigated, preventing people from using scheme water reticulation on lawn or gardens on more than two prescribed days a week.

While the water restrictions saved a considerable amount of water – 50 gigalitres (out of a normal usage of just over 300 gigalitres) – the restrictions were not sufficient to ensure environmental conditions would not continue to be breached.

However, economic and social values were also considered when the decision was made to impose the water restrictions. Stricter scheme water restrictions, for example a full sprinkler ban or even a ban on all outside water use, may have been of more benefit to the environment, but would have had significant social and economic impacts. For example, many garden supply centres and lawnmowing services could have gone out of business. It was estimated that a total sprinkler ban could be a net economic loss in the gardening industry of Perth of over $400 million a year (ECS, 2002).[1]

22.2.3 Environmental water use

Changes in both the quantity and quality of water can impact on the environment and its use of water.

Human activities have had a significant effect on the natural water cycle and water flows. In riverine systems, abstraction and the creation of dams or weirs can alter the quantity and delivery of river flow. This can impact on channel form. Reduced flow, for example, may cause the build up of sediment. There may also be less turbulence, which has the potential of deoxygenating the water and can cause fish deaths.

Natural fluctuations in water levels, turbulence and water speed in rivers are crucial to both flora and fauna, influencing zonation, migration, breeding and other life-cycle processes. The regulation of river flows, which often occurs as part of either a flood protection plan or for abstraction, is a major cause of degradation of aquatic ecosystems. Many fish species have migratory habits, moving and spending most of their lives downstream but moving upstream to breed. Weirs or dams can prevent this migration and either adversely impact on breeding patterns or, if the fish breed further downstream, may increase the risk of predators.

Regulated rivers, as they result in more constant flows, can favour pest species, such as the mosquito fish that prefers the stagnant pool waters created by dams and weirs. In Western Australia, for example, floods and drying out are important natural cycles of a number of streams, billabongs and wetlands, and are vital to the life-cycle of many native species as well as helping eliminate alien species. Many river reaches are also now isolated from their floodplains, which has removed an important spawning area for fish as well as the primary source of organic matter required for productivity.

Water use and the associated infrastructure can cause major changes to stream flow, such as altering seasonality of flows, rapid changes in water levels and temperature during releases, less frequent floods and the loss of river pools and other habitats. This can have an effect on channel stability causing slumping, erosion and sedimentation. Changes to stream flow can result in the loss of riparian vegetation, which is a major form of river degradation. Riparian

---

[1] This figure was calculated through consideration of loss of business opportunities (e.g. garden centres, turf manufacturers, lawnmowing businesses, swimming pool manufacturers) and included the expected gains for other businesses, such as paving companies.

vegetation can be undermined by erosion or drowned by inundation for long periods of time.

Furthermore, in regulated rivers water is often captured during the rainy season (often during winter) and then released for irrigation purposes during dry spells (generally summer). This reverses the natural flow pattern to which native species have adapted.

Besides the physical damage that water releases from dams can cause, cold water releases can cause fish deaths. In addition, "bottom water" (i.e. from the lower parts of the dam) releases from dams can be saturated with dissolved gases or contain toxic levels of hydrogen sulfide which can also result in fish kills.

Groundwater is similarly affected by human activity, particularly through over-pumping and through pollution/contaminants leaching from the surface into the aquifer. If more water is extracted from an aquifer than is recharged, the watertable is lowered. Subsidence of land can also follow as the previously saturated ground dries and shrinks.

In coastal areas, depletion of aquifers can allow saltwater intrusion, destroying any opportunity from further extraction of that water for potable consumption.

The loss of water from the aquifers and changes to the watertable can also have devastating effects on ecosystems reliant on certain groundwater levels. Hatton and Evans (1998) have described a number of ecosystem types that are wholly, or partially, dependent on groundwater for their continued survival. These include:

- River base flows – where they exist – are by and large wholly dependent on groundwater discharge.
- Terrestrial vegetation – wholly or partially dependent on groundwater availability. These include some tropical paperbark forests, jarrah forests, swamp sclerophyll forests, some coastal banksia and casuarina woodlands, coastal heathland communities and saline discharge samphire shrublands.
- Wetland communities, including numerous coastal wetland, mangrove and saltmarsh communities, arid-region waterholes and some swamp sedgelands and grasslands.
- Karstic cave and sinkhole ecosystems, including their "troglodyte" fauna.
- Aquifer ecosystems – poorly-known communities of bacteria and primitive invertebrates living between wetted subsoil particles in aquifers (Hatton & Evans, 1998).

Other land use and land management practices can significantly affect water resources in different ways. In the case study above, the problems of declining groundwater levels in Perth, Western Australia, was discussed. While climatic conditions, resulting in increased abstraction, has been a major contributor to the declining water levels, recent work has shown that the 23,000 hectares of commercial pine plantation on the groundwater mound is also significantly impacting on the groundwater levels as the trees consume much more water than native vegetation. Plans are in place to remove all the pine trees over the next 20 years in a bid to help restore some equilibrium to the groundwater levels (Welker Environmental Consultancy, 2002).

Environmental water use is all about the wise use of water and the sustainable management of the values or beneficial uses derived from those water resources. There are a range of environmental water use objectives, which are important considerations in terms of water auditing:

- To manage water balance
- To maintain and enhance water quality
- To encourage water conservation
- To maintain water related environmental values
- To maintain water related recreational and cultural values.

The ways in which these can be achieved will be further explored below.

22.2.3.1 Case study – Dryland Salinity

Groundwater levels in and near Perth, Western Australia, are declining. One hundred kilometres inland from Perth the story is very different. The Western Australian wheatbelt, approximately 18 million hectares in size covering much of the south-west corner of Australia, is facing a huge environmental challenge in the form of dryland salinity. While Australia is, with the exception of the Antarctic, the driest continent on the earth, the Western Australian wheatbelt suffers from rising groundwater levels due to extensive clearing of land for agriculture. Annual crops do not take up as much water as native vegetation and this has created a problem of excess recharge in a geographically ancient and ill-drained, flat environment. Salt accumulated in the soil, on average 2000 tonnes per hectare, over thousands of years is being brought to the surface through rising groundwater levels, reducing or destroying the agricultural productive capacity of the land. Where the saline water can drain, it inevitably ends up in wetlands or rivers, increasing the salinity levels of these resources and endangering native flora and fauna. It has been estimated that up to 450 plant species endemic to the region risk extinction (Government of Western Australia, 2000).

At present around two million hectares, or just over 10 per cent of the area, are salt-affected with estimations of the potential area at risk of salinity over the next 100 years being around six million hectares, or over 30 per cent of the area. It has also been estimated that over $100 million a year, over a 30-year period, is required to reduce the area potentially at risk (Government of Western Australia, 2000). There are a range of management options available, such as revegetation, particularly in recharge areas to lower groundwater levels, the introduction of high water using perennial plants and sub-surface drainage. However, all options have a high economic cost that landholders find difficult to meet with success not guaranteed, given the huge areas involved. Sub-surface drainage, for example, is not only costly but there are major difficulties with discharge as the land is so flat. The most economically cost-effective options from the landholders' perspective is to drain the saline (and generally highly acidic) water into water courses and wetlands. However, this can have major environmental consequences and is generally not permitted.

22.2.4 Water in the urban environment

Urbanisation has a considerable hydrological impact, particularly on flood runoff. Research in a range of countries, including the United States and Britain, has shown that flood runoff increases in urban areas due to the extended impermeable surfaces of bitumen, tiles and concrete (Goudie, 1986).

The installation of sewers and storm drains tends to accelerate runoff due to more efficient conveyance systems, which may also be increased in urban areas due to low vegetation densities leading to limited evapotranspiration.

Stormwater runoff from urban areas can contain large amounts of contaminants, derived from litter, garbage, fertilisers, pesticides and herbicides, vehicle drippings, industry, construction, animal droppings, detergents (e.g. from washing cars) and the chemicals used for snow and ice clearance. The contaminants come not only from roads, but pathways, roofs and gardens.

> 'In New York City, some half million dogs leave up to 20,000 tonnes of pollutant faeces and up to 3.8 million litres of urine in the city streets each year, all of which is flushed by gutters to stormwater sewers.'
>
> (Goudie, 1987:181).

Stormwater ends up somewhere – often in receiving waterbodies such as rivers, lakes, wetlands, estuaries or coastal waters. While receiving water bodies can absorb some level of contaminants, beyond a certain point they can cause major environmental damage.

In fact, stormwater is the single most important source of pollution of water bodies in and around cities. These impacts include:

- Increases in peak discharge and frequency affecting channel scouring
- Increases in nutrient loads with associated risk of eutrophication and impacting on primary production as a result of nuisance algae

- Increases in toxicants (heavy metals, pesticides, ammonia) accumulating in estuarine and near shore sediments and impacting on aquatic biota
- Increases in sediment and suspended solids
- Oxygen demanding substances impacting on dissolved oxygen (depleting availability to aquatic animals) and sediment redox processes (reducing water quality by causing the release of pollutants bound in sediments)
- Bacteria impacting on recreation and water supply suitability
- Rubbish and debris impacting on visual quality of waterways, and
- Reduction in biodiversity within the waterways and increased risk of algal blooms and other ecological perturbations (Commonwealth of Australia, 2002:158).

An additional problem with water use in the urban environment is that of water efficiency, or the lack of it. Worldwide, 50 litres per person per day has been determined as the basic water requirement for drinking, sanitation, bathing and cooking. People in Asia, Africa and Latin America use 50-100 litres per day. This can be compared to the United States, where people now use up to 400-500 litres per day (NLWRA, 2000). During the 1900s, a time of rapid urbanisation, per capita water usage in the USA grew by 500 to 800 per cent (Anderson, 1995).

In Australia, total water consumption, and per capita consumption, is trending upward, although the influence of climate variability is a confounding factor over the short term. Recent droughts across the continent have affected the way in which people use water, mostly due to imposed restrictions. However, in Australian cities, efficient water use is still perceived as an emergency measure to be adopted during drought conditions (Commonwealth of Australia, 2000).

In many parts of the world inefficient water use prevails because consumers do not have to pay the full cost of the water they use. Current government policies do not adequately reflect the costs of operation, maintenance and capital for water projects and fail to account for the environmental damage associated with water storage and delivery (Anderson, 1995).

As a result, in many places, such as Libya mentioned in the introduction, above, are using much more water than is sustainable.

> 'We can blame the policy makers of the last century, as year after year, water authorities dished out more water than there was to go around, mainly through ignorance but also in response to greed and cronyism.'
>
> (Fullerton, 2001:4).

In Australia, the above problems are starting to be addressed through the adoption of Water Sensitive Urban Design (WSUD) principles. When WSUD was initiated in Australia in the early 1990s, it was heralded as a new approach to urban planning and design that integrated flood control (water quantity) and nutrient control (water quality). The intention was to move from the traditional "conveyance" approach to stormwater management, which aimed to get the water into pipes and into receiving water bodies as quickly as possible, to a "treatment train" approach. A treatment train consists of a number of water quality management measures that may be implemented in series or concurrently to manage stormwater. This includes:

- slowing the rate at which the water moves
- encouraging infiltration
- using natural systems to convey stormwater
- trapping the "first flush" (runoff from the first rain event following a long dry spell) of stormwater runoff which contains disproportionately large amounts of pollutants, and
- detaining stormwater in natural systems for sufficient periods to allow settling and biological processes to remove pollutants prior to discharge to natural ecosystems.

Water Sensitive Urban Design aims to bring stormwater management out of pipes in the ground and to make the entire stormwater treatment network part of the urban fabric through the use of multiple use corridors and Best Management Practice treatment trains. Vegetated swales, filter strips, extended detention basins and constructed wetlands are all part of fully functioning stormwater treatment systems and may also serve other uses (Water and Rivers Commission, 1999:3).

Stormwater should be considered as an asset – a resource that can be used – rather than a nuisance to be disposed of quickly. The WSUD philosophy is for stormwater to be managed as part of the global water balance, with the principles being:

- Incorporation of water resource issues early in the land use planning process
- Addressing water resource management at the catchment and subcatchment level
- Recognising stormwater management to be part of total water-cycle management
- Wherever possible, using the natural contours to incorporate and enhance functions of the natural stormwater system
- Maximising local on-site storage and utilising natural runoff channels

- Emphasising the use of vegetation (particularly indigenous vegetation) in stormwater management to promote filtering and slowing of runoff to maximise settling of particulate-bound pollutants
- The designation of water sensitive design precincts which delineate land suitable for development from that to be set aside for stormwater management, and the protection of water resources including protection of water related environmental, cultural and recreational values,
- The opportunity to create a linked open space system which recognises the natural drainage function, and
- Water sensitive urban (residential) design approach to road layout, lot layout (cluster housing) and streetscape (Water and Rivers Commission, 1999:9).

22.2.5 Management techniques

In section 22.2.3 above, the environmental water use objectives were briefly discussed. This section will outline management techniques to address each of these objectives and will give some idea of the potential role that water auditing can have in monitoring and assessing the efficacy of these techniques.

22.2.5.1 Managing water balance

Water balance should be managed to ensure wetland and aquifer levels, recharge and stream flow are maintained, that flood damage is prevented or minimised and to prevent excessive erosion of waterways slopes and banks.

Watertables often rise in urban areas due to reduced levels of vegetation and greater areas of impervious surfaces. The traditional approach to deal with this water balance problem has been to drain the area.

Drains need to be set at a level that will help reduce the risk of flooding, but not so low as to lower the watertable below what it would be in its natural state. One approach is to ensure that the drain invert is not set below the Annual Average Maximum Groundwater Level (AAMGL).

Water balance can also be managed through ensuring that wetlands and water courses have adequate vegetated buffers, as vegetation can help regulate groundwater levels.

WSUD encourages the use of "multiple use corridors" as part of stormwater management. These corridors can be of any width, and can be planned for a variety of uses, for example playgrounds or landscaped public open space. These can help maintain water balance as well.

Any urban water management system must provide adequate provision for infiltration and adequate floodwater escape routes.

22.2.5.2 *Maintain and enhance water quality*

Water quality needs to be managed in the urban environment to minimise water borne sediment loading, to protect riparian and fringing vegetation and to minimise the export of pollutants to surface and groundwater.

This can be achieved by ensuring:

- An adequate concept design is developed for pollutant and nutrient management,
- That adequate land area is provided for water quality management,
- That source control, for example the reduction of inputs such as fertiliser use, and treatment trains are used in stormwater management to control water quality,
- That drain inverts are not set below the AAMGL, as excessive drainage can increase the concentration of pollutants discharged into receiving water bodies, and
- There are adequate buffer zones and fringing vegetation is protected.

One of the major problems in dealing with stormwater pollution can be the lack of clarity in management arrangements. Pollution/contaminants can enter a water source many kilometres away from where the problems actually surface and the agency with the responsibility for maintaining water quality in the receiving water body may have no control over the quality of the stormwater that flows into the water body.

22.2.5.3 *Encourage water conservation*

Given the increasing pressure on water resources, any means to reduce the overall per capita use of water should be encouraged. This is particularly important in countries of limited water resources, such as Australia, where efficient water use should be the norm, not the exception.

An additional issue is the use of highly treated water. In Perth, Western Australia, approximately 50 per cent of the scheme water (public water supply) goes straight onto gardens. The scheme water is treated to appropriate drinking water standards, yet only about one per cent is actually consumed by humans (Loh *et al.*, 2002).

A major water conservation objective should be, therefore, to protect and minimise the import and use of high quality scheme water. This can be partly achieved through promoting the reuse of stormwater. In Perth, homeowners are generally encouraged to install garden bores as this reduces pressure on the scheme water and, in addition, helps to regulate the groundwater levels in the superficial aquifer.

Wastewater can often be treated and recycled as well, which helps encourage water conservation. Irrigation methods are improving all the time, for example the use of drip feeders, not irrigating during the day (when evaporation rates are highest) and using mulch.

22.2.5.4 Maintain water related environmental values

Water-related environmental values can be maintained through appropriate planning, to ensure the environmental values are protected. For example, wetlands and waterways need adequate buffers. Buffer distances will vary depending on the area and the proposed land use. For example, in Western Australia a general rule of thumb is to have a minimum of 50 meter buffer from the edge of the wetland-dependant vegetation in residential areas, with the buffer requirements increasing to 200 metres for some industries.

It is important to ensure that the "first flush" is trapped and not disposed of in receiving water bodies. As for other objectives, water balance should be maintained and water quality managed.

Another way of maintaining water related environmental values is to ensure an appropriate environmental flow. An environmental flow is produced by releasing dam or bore water to simulate, maintain or repair natural flow patterns when an environment has been highly modified, for example through the presence of a dam or weir.

In highly modified environments, which occur almost anywhere where humans are taking and using water, it is impossible to restore flows to pre-regulation conditions. Rather, the aim of environmental flow management is to mimic natural flow regimes, providing cues for key lifecycle events such as spawning and migration. It can also rehabilitate and improve ecosystems.

The concept of an environmental flow includes managing for:

- Water level rise and fall
- Duration of a flow event
- Velocity of water in the channel
- Seasonality of flows
- Need for flood pulses or high flows, and
- Protection of water levels during periods of low or no flow.

This can be achieved through modifying existing releases, increasing water use efficiency, restricting diversions and extraction, and by increasing allocations from large storage areas for the environment.

The concept of environmental flow forms part of the consideration of ecological water requirements. This concept, as shown above, can be readily understood within the context of surface water, but groundwater ecosystems have certain water requirements as well, which can take the form of certain minimum and/or maximum water levels at certain times of the year or certain water quality conditions. In modified environments, for example where groundwater abstraction is taking place or where urbanisation has raised groundwater levels, ecosystem reliant on specific groundwater conditions can be artificially maintained, for example through pumping either into or out of the system.

22.2.5.5 Maintain water related recreational and cultural values

The final environmental water use objective recognises that humans are also part of the environment and that water sources can be important areas for recreation and/or have high cultural value, particularly for indigenous populations. These values need to be recognised in planning and maintained wherever possible.

22.2.6 Conclusion

This section has explored issues associated with environmental water use and water resource management in the urban environment. From an auditing perspective, it is important to recognise the objectives for environmental water use and how these objectives can be achieved through appropriate management techniques.

Throughout the world today, water is generally not being used wisely or sustainably and the environment is not a primary consideration. However, unless some balance is restored in the way in which we use and manage water, the pressure on and competition for our limited water resources will only increase.

To aid in a semi-quantitative assessment of the many factors involved in water resource management, environmental water use and water sensitive design, Appendix 22.1 contains a series of check-sheets for the practitioner, courtesy of the Water and Rivers Commission of Western Australia.

22.3 DESALINATION OF WATER RESOURCES

(Note units used in this section are commonly used in the field but are not always SI units)

22.3.1Desalination Plants Worldwide

Of the more than 7 500 desalination plants in operation worldwide, 60% are located in the Middle East. The world's largest plant in Saudi Arabia produces

484 million litres per day (MLD) of desalted water. In contrast, 12% of the world's capacity is produced in the Americas, with most of the plants located in the Caribbean and Florida. To date, only a limited number of desalination plants have been built along the California coast, primarily because the cost of desalination is generally higher than the costs of other water supply alternatives available in California (e.g. water transfers and groundwater pumping).

22.3.2 Desalination Technologies

Desalination is a process that removes dissolved minerals (including but not limited to salt) from seawater, brackish water, or treated wastewater. A number of technologies have been developed for desalination, including reverse osmosis (RO), distillation, electrodialysis, and vacuum freezing. Two of these technologies, RO and distillation, are being considered by municipalities, water districts, and private companies for development of seawater desalination in California. These methods are described below.

22.3.2.1 Reverse Osmosis (RO)

In RO, feedwater is pumped at high pressure through permeable membranes, separating salts from the water. The feedwater is pre-treated to remove particles that would clog the membranes. The quality of the water produced depends on the pressure, the concentration of salts in the feedwater, and the salt permeation constant of the membranes. Product water quality can be improved by adding a second pass of membranes, whereby product water from the first pass is fed to the second pass.

22.3.2.2 Distillation

In the distillation process, feedwater is heated and then evaporated to separate out dissolved minerals. The most common methods of distillation include multistage flash (MSF), multiple effect distillation (MED), and vapour compression (VC).

In MSF, the feedwater is heated and the pressure is lowered, so the water "flashes" into steam. This process constitutes one stage of a number of stages in series, each of which is at a lower pressure. In MED, the feedwater passes through a number of evaporators in series, Vapour from one series is subsequently used to evaporate water in the next series, The VC process involves evaporating the feedwater, compressing the vapour, then using the heated compressed vapour as a heat source to evaporate additional feedwater.

Some distillation plants are a hybrid of two or more technologies. The waste product from these processes is a solution with high salt concentration.

22.3.3 Input Water (Feedwater)

Desalination plants may use seawater (directly from the ocean through offshore intakes and pipelines, or from wells located on the beach or seafloor), brackish groundwater, or reclaimed water as feedwater. Since brackish water has a lower salt concentration, the cost of desalting brackish water is generally less than the cost of desalting seawater. Intake pipes for desalination plants should be located away from sewage treatment plant outfalls to prevent intake of discharged effluent. If sewage treatment discharges or other types of pollutants are included in the intake, however, the pre- and post-treatment processes should remove the pollutants.

22.3.4 Product Water

Distillation plants produce a high-quality product water that ranges from 1.0 to 50 parts per million (ppm) total dissolved solids (TDS), while RO plants produce a product water that ranges typically from 10 to 500 ppm TDS. The recommended California drinking water standard for maximum is 500 mg/L TDS, which is equivalent to 500 ppm. In desalination plants that produce water for domestic use, post-treatment processes are often employed to ensure that product water meets the health standards for drinking water as well as recommended aesthetic and anti-corrosive standards. Desalination product water may be used in its pure form (e.g. for make-up water in power plant boilers) or it may be mixed with less pure water and used for drinking water, irrigation, or other uses. The desalinated product water is usually more pure than drinking water standards, so when product water is intended for municipal use, it may be mixed with water that contains higher levels of total dissolved solids. Pure desalinated water is highly acidic and is thus corrosive to pipes, so it has to be mixed with other sources of water that are piped onsite or else adjusted for pH, hardness, and alkalinity before being piped offsite.

22.3.5 Product Water Recovery

The product water recovery relative to input water flow is 15 to 50% for most seawater desalination plants. For every 100 gallons of seawater, 15 to 50 gallons of pure water would be produced along with brine water containing dissolved solids. A desalination plant's recovery varies, in part because the particulars of plant operations depend on site-specific conditions. In several locations in California, pilot projects are being proposed to test plant operations before full-scale projects are built.

22.3.6 Pre-treatment Processes

Pre-treatment processes are needed to remove substances that would interfere with the desalting process. Algae and bacteria can grow in both RO and distillation plants, so a biocide (usually less than 1 mg/L of chlorine) is required to clean the system. Some RO membranes cannot tolerate chlorine, so dechlorination techniques are required, Ozone or ultraviolet light may also be used to remove marine organisms. If ozone is used, it must be removed with chemicals before reaching the membranes. A RO technology has been developed recently that does not require chemical pre-treatment. In RO plants, suspended solids and other particles in the feedwater must be removed to reduce fouling of the membranes. Suspended solids are removed with coagulation and filtration. Metals in the feedwater are rejected along with the salts by the membranes and are discharged in the brine. With normal concentrations for metals in seawater, the metals present in the brine discharge, though concentrated by the RO process, would not exceed discharge limits. Some distillation plants may also need to remove metals due to potential corrosion problems.

22.3.7 Filter Backwashing, Membrane Cleaning and Storage, Scaling Prevention and Removal, and Pipeline Cleaning

The filters for pre-treatment of feedwater at RO plants must be cleaned every few days (backwashed) to clear accumulated sand and solids. The RO membranes must be cleaned approximately four times a year and must be replaced every three to five years. Alkaline cleaners are used to remove organic fouling, and acid cleaners are used to remove scale and other inorganic precipitates. All or a portion of RO plants must be shut down when the membranes are replaced. When RO plants are not used continuously, the RO membranes must be stored in a chemical disinfection/preservation solution that must be disposed of after use. Distillation plants can also be shut down for tube bundle replacement, which is analogous to membrane replacement.

Desalination plant components must be cleaned to reduce scaling - a condition where salts deposit on plant surfaces, such as pipes, tubing or membranes. Scaling is caused by the high salt concentration of seawater and can result in reduced plant efficiency and corrosion of the pipes. In general, scaling increases as temperature increases; thus scaling is of greater concern for distillation plants, since RO plants require lower temperatures to operate. Introducing additives to inhibit crystal growth, reducing temperature and/or salt concentrations, removing scale-forming constituents, or seeding to form particles, can reduce scaling. Once scales have formed, they can be removed with chemical or mechanical means. In addition to scaling, both RO and distillation plant intake and outfall structures and pipelines can become fouled with naturally occurring organisms or corroded.

Structures and pipelines may be cleaned by mechanical means or by applying chemicals or heat. Feedwater may also be de-aerated to reduce corrosion.

22.3.8 Waste Discharges

Desalination plants produce liquid wastes that may contain all or some of the following constituents: high salt concentrations, chemicals used during de-fouling of plant equipment and pre-treatment, and toxic metals (which are most likely to be present if the discharge water was in contact with metallic materials used in construction of the plant facilities). Liquid wastes may be discharged directly into the ocean, combined with other discharges (e.g., power plant cooling water or sewage treatment plant effluent) before ocean discharge, discharged into a sewer for treatment in a sewage treatment plant, or dried out and disposed of in a landfill. Desalination plants also produce a small amount of solid waste (e.g., spent pre-treatment filters and solid particles that are filtered out in the pre-treatment process). For example, the capacity of the City of Santa Barbara's desalination plant is 7 500 AF/yr, about 27 megalitres per day (MLD). In May 1992, the plant produced 25.4 MLD of product water and generated 31 MLD of waste brine with a salinity approximately 1.8 times that of seawater. An additional 6.4 MLD of brine was generated from filter backwash. Assuming that concentrations of suspended solids in the seawater feed range from 10 to 50 ppm, approximately 1.2 to 3.5 cubic metres per day of solids were generated (Woodward-Clyde, 1991).

22.3.9 Energy Use

The energy used in the desalination process is primarily electricity and heat. Energy requirements for desalination plants depend on the salinity and temperature of the feedwater, the quality of the water produced, and the desalting technology used. Estimates for electricity use requirements for various technologies for seawater desalination are: MSF, less than 4.5 kWh/m^3; RO, sea water approx. 6.5 kWh/m^3, brackish water approx. 2.5 kWh/m^3 or less.

In addition to electricity requirements, Multi Stage Flash (MSF) and some VC plants also use thermal energy to heat feedwater. Because of the inefficiency of converting thermal energy to electricity, there is a high energy 'penalty' if electricity is used to heat feedwater. For example, in addition to the 3 500 to 7 000 kWh/AF of energy required for electricity, the thermal energy needs for a MSF distillation plant is estimated at 270 million BTU/AF (about 26,000 kWh/AF); for IVIED plants, the estimated additional thermal energy requirements are 230 million BTU/AF (about 22 000 kW/AF). Consequently, the total energy needs for distillation technologies are higher than for RO technologies. Energy use requirements for desalination plants are high. For example, an estimated 50 million kWh/yr would be required for full-time operation of the City of Santa Barbara's desalination plant to produce 7 500 AF/yr of water. In contrast, the energy needed to pump 7 500 AF/yr of water from the Colorado River Aqueduct or the State Water Project to the Metropolitan Water District (MWD) of Southern California is 15 to 26 million kWh/yr. These energy requirements may be compared to the energy use of a small to medium sized industrial facility (such as a large refinery, small steel mill, or large computer centre) which uses 75 000 to 100 000 kWh/yr. Both RO and distillation plants can benefit from co-generation plants to reduce energy use. Since increased energy use may cause adverse environmental impacts, the individual and cumulative impacts of energy use and production at a proposed desalination plant will require case-by-case analysis.

22.3.10 Comparison of Distillation and Reverse Osmosis Technologies

One advantage of distillation plants is that there is a greater potential for economies of scale. Distillation plants also do not shut down a portion of their operations for cleaning or replacement of equipment as often as RO plants, although distillation plants can and have shut down for tube bundle replacement and cleaning. Pre-treatment requirements are greater for RO plants, because coagulants are needed to settle out particles before water passes through the membranes. Unlike RO plants, distillation plants do not generate waste from backwash of pretreatment filters. Advantages of RO plants over distillation include: RO plant feedwater generally does not require heating, so the thermal impacts of discharges are lower; RO plants have fewer problems with corrosion; RO plants usually have lower energy requirements; RO plants tend to have higher recovery rates - about 45% for seawater; the RO process can remove unwanted contaminants, such as trihalomethane-precursors, pesticides, and bacteria; and RO plants take up less surface area than distillation plants for the same amount of water production.

22.3.11 Costs of Desalinated Water

The cost to produce water from desalination depends on the technology used and the plant capacity, among other factors. For example, the cost of desalted water in Santa Barbara ($1 900/AF) results from the following: a write-off of the capital cost over a short five-year period, high financing costs, and high energy costs. The overall costs of water production are about the same for RO and some forms of distillation plants. Price estimates of water produced by desalination plants in California range from $1 000 to $4 000/AF. For long-term projects, capital costs would most likely be amortized over an assumed plant life of 20 to 30 years. Capital costs for RO plants tend to be lower than for distillation plants. Operating a plant on a part-time, rather than full-time, basis may be more expensive in the long run because maintenance and capital costs must be paid while the plant is shut down.

22.3.12 Costs of Other Water Sources

A number of California coastal communities are facing water shortages. Although the communities may have relatively inexpensive existing supplies of water, the supplies are perceived as being insufficient to meet community needs. New water supplies are more expensive than existing supplies, and in some cases the prices are comparable to desalinated water. In 1991, the Metropolitan Water District (MWD) of Southern California paid $27/AF for water from the Colorado River and $195/AF for water from the California Water Project. New sources of water would have cost $128/AF from the Imperial Irrigation District and $93/AF from Arvin Edison Water Storage in Kern County (if water was available during the drought) (Muir, 1991). Non-interruptible untreated water for domestic uses in San Diego is purchased from the MWD for $269/AF; treated water costs an additional $53/AF. The least expensive new supplies, other than desalination, would cost $600-$700/AF (Hess, 1991; Yamada, 1992). In Santa Barbara, untreated water from the Cachuma reservoir costs $35/AF. Development of new wells to use further the City's groundwater basins would cost $200/AF, while new groundwater wells in the mountains would cost approximately $600-700/AF. Enlarging Cachuma Reservoir, if feasible, is estimated to cost $950/AF. During the recent drought, the City purchased water from the State Water Project on a temporary, emergency basis at a cost of $2 300/AF. This water was made available through a series of exchange arrangements with water agencies between Santa Barbara and the MWD. Permanently tying into the State Water Project is estimated to cost $1 300/AF (Brown, 1992).

22.4 REFERENCES

Anderson, Terry L. (1995). "Water options for the blue planet", *The True State of the Planet*, Ronald Bailey (Ed), The Free Press, New York, pp.267-294.

Brown, D (1992) City of Santa Barbara. *Pers. Comm.*

Bruntland, Gro Harlem (1987). *Our Common Future*, United Nations, New York.

Commonwealth of Australia (2002). *The Value of Water*. Inquiry into Australia's Urban Water Management. Report of the Senate Environment, Communications, Information Technology and the Arts References Committee, Canberra.

Construction Industry Research and Information Association (CIRIA) 1999. *Sustainable urban drainage systems design manual for Scotland and Northern Ireland*, CIRIA, London.

Economics Consulting Services (ECS) (2002). *Residential Water Restrictions – Economic Impact Assessment*, report prepared for the Water and Rivers Commission, Perth.

Fullerton, Ticky (2001). *Watershed*, ABC Books, Sydney.

Goudie, Andrew (1986). *The Human Impact on the Natural Environment* 2nd Edition, Basil Blackwell, Oxford.

Government of Western Australia (2000). *The Salinity Strategy*, Perth.

Hatton, T.J. and Evans, R. (1998). "*Dependence of Ecosystems on Groundwater and its significance to Australia*", Canberra, ACT. Land & Water Resources Research and Development Corporation, Occasional Paper 12/98.

Hess, G (1991) San Diego County Water Authority (SDCWA), *Pers Comm*

Loh, M., L. Gould, P. Coghlan, C. Jeevaraj & G. Hughes (2002). Domestic water use study – the next step forward, in *Proceedings of 27th Hydrology & Water Resources Symposium*, 20-23 May.

Lowe, Ian (2002). *Towards Sustainability*, unpublished (Perth, Australia).

Muir, R (1991) Metropolitan Water Department (MWD), *Pers. Comm.*.

National Land and Water Resources Audit (NLWRA) 2000. *Audit for year 2000.* (http:/www.nlwra.gov.au)

Smith, David Ingle (1998). *Water in Australia, Resources and Management*, Oxford University Press, Australia.

UNEP-International Environmental Technology Centre (2002). *Environmentally sound technologies for wastewater and stormwater management. An international resource book.* UNEP-IETC Technical publications Services No. 15, IWA Publications, London.

Water and Rivers Commission (1999). *A manual for managing urban stormwater quality in Western Australia*, Government of Western Australia, Perth.

Welker Environmental Consultancy (2002). *Section 46 Review of Environmental Conditions on Management of the Gnangara and Jandakot Mounds, Progress Report 2002* (prepared for the Water and Rivers Commission, Perth).

Woodward-Clyde Consultants (1991) EIR for the City of Santa Barbara and Ionics Inc.'s Temporary Desalination Project.

Yamada, R (1992) San Diego County Water Authority (SDCWA), *Pers Comm.*

22.5 BIBLIOGRAPHY

Western Australia. Water Corporation. (2000) *Desalination: a viable resource : a strategic review of desalination use in Western Australia.* Water Corporation, Leederville, W.A.

Desalination. (Journal) Amsterdam : Elsevier Scientific Pub. Co.
http://www.desalination.com/Desalination.html

APPENDIX 22.1 ENVIRONMENTAL WATER USE AND WATER SENSITIVE DESIGN CHECKLIST

DATE:
ASSESSMENT DONE BY:
NUMBER OF HOME UNITS:
LOCATION OF SITE:

LEVEL OF APPLICATION OF WATER SENSITIVE URBAN DESIGN (circle appropriate level):

Strategic Subdivision Catchment
Rezoning Houselot Sub-
catchment

DESCRIPTION OF SITE OR AREA:

Catchment: Depth to Groundwater:
Groundwater Area: % of Area Seasonally Waterlogged:
Soil type: Natural Wetlands (% coverage):
Wetland Significance(Int, Nat State, Reg, Local)
Stratigraphy: % Remnant Veg. Present
Development zoning: Sewerage system:
Stormwater system: Land Capability Assessment:

Part One: Regional Water Resources Imperatives

HAVE THE REGIONAL AND LOCAL WATER RESOURCE MANAGEMENT OBJECTIVES FOR THE AREA BEEN MET?

Statutory Policies, Ordinances, Bylaws

eg. Check Peel Harvey Catchment Environmental Protection Policy (EPP), Peel Harvey Catchment Statement of Planning Policy; Swan Canning Rivers EPP; Swan Coastal Plain Wetlands EPP; South West Agricultural Zone Wetlands EPP; Priority 1, 2, 3, Water Supply Catchment Areas.

Water Resources Management Plans

eg. Check Perth Bunbury Water Allocation Studies; Wetland Management Objectives - Hill et. al. (1996); Middle Canning Catchment Water Resources Management Plan (Evangelisti & Associates, 1995); Urban Water Management in the South East Corridor (Landvision et al 1997); Strategic Drainage Plans (Evangelisti & Assoc, 1994, 1996; Tan and Loh, 1989); Average Annual Maximum Groundwater Level Advice (for 1.2, 1.5m pad height clearance & as the minimum level for drainage inverts (Rockwater, 1997); Major River River Floodplain Management Advice

Does the proposal protect Regional and Local Water Resource Management Objectives?

WATER RESOURCE MANAGEMENT CRITERIA	YES/ NO	COMMENTS
WATER BALANCE		
Maintained natural hydrologic processes, water regimes and drain inverts set no lower than the Annual Average Maximum Groundwater Level (AAMGL)		
Adequate protection of natural watercourses/wetlands		
Are stormwater management corridors incorporated into landscaped public open space?		
Are there adequate infiltration and recharge basins?		
Is an adequate flood escape route provided?		
Is the road layout water sensitive?		
WATER QUALITY		
Adequate protection of natural wetlands and augmenting with adequate concept plan design for artificial nutrient stripping basins and artificial wetlands to prevent the export of nutrients.		
Provision of an adequate number of water quality basins		
International best practice erosion and sediment control		
Is there evidence of a "treatment train" to control water quality?		
Drain inverts set no lower than the Annual Average Maximum Groundwater Level		
Are there adequate buffer zones and fringing vegetation?		
Saline intrusion and Marine sediments with Acid and Heavy metal Pollution risks to ground water competently considered and avoided		
WATER CONSERVATION		
Development outside of Priority 1 Catchment areas		
Development outside Priority 2 Catchment areas or compatible development within		
Development outside Priority 3 Catchment areas or compatible development within		
Is private bore use and self supply of water being encouraged?		
Is water sensitive landscaping being encouraged?		
Is water efficient reticulation technology used		
WATER RELATED ENVIRONMENTAL, RECREATIONAL, AND CULTURAL VALUES		
Adequate protection of recognised "conservation" and "resource enhancement" wetlands or vegetated waterways: maintaining their functions, uses and attributes		
Appropriate allocation of public open space		

Adequate linkage of public open space and areas of conservation open space using multiple use corridors.		
Adequate protection of Aboriginal or historical sites		
Provision of water dependant and water orientated recreational opportunities		
Are the life cycles of Aquatic fauna being adequately protected		
Are the natural characteristics of artificial drainage channels being managed or restored?		
Are large box culverts used on roads to allow fauna movement along and adjacent to waterway		
Is catchment hydrological processes and remnant vegetation particularly on recharge areas and foreshores being retained and protected sufficiently?		

Score *

Overall, is the site a successful example of water sensitive urban design? (If not, why not?)

Part Two : Design Response and Assessment of Planning and Management Practices at this Site

FOR WATER BALANCE: (to maintain water regimes and manage water balance)

MANAGEMENT PRACTICE	IS IT USED? (Y/N)	MANAGEMENT PRACTICE	IS IT USED? (Y/N)
Water sensitive stormwater management system		Parking lot storage	
Dry detention basin		Porous pavements	
Infiltration retention basin		Rooftop run-off disposal	
Infiltration trench		Swales	
Flush/strip kerbing		Urban forestry or revegetation	
Modular paving		Wet detention basin	
Overland flow/filter strips			

Score *

* Give each table a generalised score out of 100 relating to how adequately the objectives or management practices of water sensitive urban design have been met or planned at the site.

TO PREVENT SCOURING, EROSION, AND SEDIMENTATION:
(to ensure water quality)

MANAGEMENT PRACTICE	IS IT USED? (Y/N)	MANAGEMENT PRACTICE	IS IT USED? (Y/N)
Water sensitive stormwater management system		Preservation of Rem Veg	
Aprons		Rip-rap	
check dams		Seeding and revegetation	
Cribbing/deflectors		Slope configurations	
Diversion		Stilling basins	
Drop structures		Stone center drains	
Energy dissipaters		Stream rehabilitation	
Gabions/mattresses		Top soiling	
Hydroseeding		Turfing	
Mulching		Erosion and Sediment Control Best Practice	

Score *

TO PREVENT STORMWATER RUN-OFF POLLUTION: (to ensure water quality)

MANAGEMENT PRACTICE	IS IT USED? (Y/N)	MANAGEMENT PRACTICE	IS IT USED? (Y/N)
Water sensitive stormwater management system		Preservation of remnant vegetation - vegetation filters	
Alum treatment		Oil/grit separators	
Artificial wetlands		Overland flow/filter strips	
Coalescing plate separators		Porous pavements	
Extended dry detention basin		Spill control separators	
Gross pollutant trap		Soil amendment	
Helical bend regulators		Swales	
Infiltration retention basin		Swirl concentrators	
Infiltration trench and bores		Urban forestry and revegetation	
Nutrient and irrigation management program/plan		Wet detention basin	
Modular pavement		Erosion and Sediment Control Best Practice	

Score *

* Give each table a generalised score out of 100 relating to how adequately the objectives or management practices of water sensitive urban design have been met or planned at the site.

FOR WATER CONSERVATION: (to reduce scheme water irrigation and the total irrigation required)

MANAGEMENT PRACTICE	IS IT USED? (Y/N)	MANAGEMENT PRACTICE	IS IT USED? (Y/N)
Dual supply - reuse of stormwater		Soil improvement	
Cluster development		Turf management	
Hydrozones		Water harvesting	
Irrigation design/m'ment		Water saving devices	
Retain remnant and existing vegetation		Windbreaks	
Self supply of water		Xeriscaping	
Soil cultivation/mulches		Trickle Irrigation	

Score *

Part Three: Environmental, Water Resources & Societal Context Summary

WHAT WATER RELATED ENVIRONMENTAL, RECREATIONAL, AND CULTURAL VALUES ARE ADEQUATELY ADDRESSED?*

Functions	**YES/NO**	**Attributes**	**YES/NO**
Nursery/breeding area		Richness/diversity of flora and fauna	
Habitat for fish/wildlife		Aesthetic qualities	
Wildlife/veg. corridor		Rare wetland type	
Maintain natural systems / recharge		Rare/endangered species	
Nutrient Stripping/Flood protection/		Rare ecosystem	
Uses		Wilderness	
Plant products		Significant gene pool	
Animal products		Spiritual/symbolic site	
Recreation		Scientific learning	
Tourism		Historical significance	
Research/monitoring site		Aboriginal significance	
Education site		Heritage site/area	

Score *

* Give each table a generalised score out of 100 relating to how adequately the objectives or management practices of water sensitive urban design have been met or planned at the site.

APPENDIX 22.2 GLOSSARY OF TERMS

(Multiply British thermal unit (BTU) values of fuel by approx. 0.33 to calculate a kWh equivalent of electricity, because the efficiency of conversion from thermal energy to electricity is about 33%. The normal *unit* conversion factors apply as well.)

Acre-foot (AF). A unit for measuring the volume of water. One acre-foot is equal to approximately 1.23 megalitres. (the volume of water that will cover one acre to a depth of one foot).

Biocide. A chemical used to kill biological organisms (e.g. chlorine).

Brine. Water that contains a high concentration of salt. Brine discharges from desalination plants may include constituents used in pretreatment processes, in addition to the high salt concentration seawater.

BTU (British Thermal Unit). A standard unit for measuring a quantity of thermal energy, either electricity, natural gas or any other source of energy. One BTU is the amount of thermal energy required to raise the temperature on one pound of water one degree Fahrenheit at sea level.

Capacity Factor. An electric utility's annual capacity factor is defined as the annual kilowatt-hour sales divided by the product of the total hours in a year and the rated capacity of the utility in kilowatts.

Coagulation. A pretreatment process used in some desalination plants. A substance (e.g. ferric chloride) is added to a solution to cause certain elements to thicken into a coherent mass, so that they may be removed.

Cogeneration. A power plant that is designed to conserve energy by using "waste heat" from generating electricity for another purpose.

De-aeration. Removal of oxygen. A pre-treatment process in desalination plants to reduce corrosion.

Distillation. A process of desalination where the intake water is heated to produce steam. The steam is then condensed to produce product water with low salt concentration.

Electrodialysis. Most of the impurities in water are present in an ionized (electrically charged) state. When an electric current is applied, the impurities migrate towards the positive and negative electrodes. The intermediate area becomes depleted of impurities and discharges a purified stream of product water. This technology is used for brackish waters but is not currently available for desalting seawater on a commercial scale.

Feedwater. Water fed to the desalination equipment. This can be source water with or without pretreatment.

Infiltration Gallery. A method used for seawater intake. Perforated pipes are arranged in a radial pattern in the sand onshore below the water level. Water in the saturated sand enters the perforated pipes.

Ion Exchange. A water treatment process. An electric charge is used to remove charged particles from solution.

Kilowatt (kW). A thousand watts. The watt is a measure of power used by electricity generating plants. One watt is equivalent to 1 Joule/second or 3.4127 BTU/hour.
Megawatt (MW). A million watts.
Microlayer. The upper few millimeters of the ocean. Fish eggs are sometimes concentrated in the microlayer.
Multiple Effect Distillation (MED). A form of distillation. Evaporators are in series, and vapor from one effect is used to evaporate water in the next lower pressure effect. This technology is in several forms, one of the most common of which is the Vertical Tube Evaporator (VTE).
Multistage Flash (MSF). A form of distillation. Intake water is heated then discharged into a chamber maintained slightly below the saturation vapor pressure of the incoming water, so that a fraction of the water content flashes into steam. The steam condenses on the exterior surface of heat transfer tubing and becomes product water. The unflashed brine enters another chamber at a lower pressure, where a portion flashes to steam. Each evaporation and condensation chamber is called a stage.
Ocean Thermal Energy Conversion (OTEC). A solar, ocean thermal desalination approach where electricity is produced by using the temperature differential between cold, deep waters and warm, shallow surface waters. Water at the ocean surface (at about 21°C) is used to heat liquid ammonia, which vaporizes at this temperature in a vacuum chamber. The ammonia vapor is used to turn a turbine to produce electricity. The vapor is then condensed by using cold water pumped up from the ocean depths (at about 2°C).
Product Water. The desalted water delivered to the water distribution system.
Reverse Osmosis (RO). A process of desalination where pressure is applied continuously to the feedwater, forcing water molecules through a semipermeable membrane. Water that passes through the membrane leaves the unit as product water; most of the dissolved impurities remain behind and are discharged in a waste stream.
Scaling. Salt deposits on the interior surfaces of a desalination plant.
Total Dissolved Solids (TDS). Total salt and calcium carbonate concentration in a sample of water, usually expressed in milligrams per liter (mg/L) or parts per million (ppm). The state-recommended Maximum Contaminant Level (MCL) drinking water standard for total dissolved solids is 500 mg/L, the upper MCL is 1 000 mg/L, and the short-term permitted level is 1 500 mg/L.
Vacuum Freezing (VF). A process of desalination where the temperature and pressure of the seawater is lowered so that the pure water forms ice crystals. The ice is then washed and melted to produce the product water. This technology is still being developed, and is not yet commercially competitive.
Vapour Compression (VC). A form of distillation. A portion of feedwater is evaporated, and the vapour is sent to a compressor. Mechanical or thermal energy is used to compress the vapour, which increases its temperature. The vapour is then condensed to form product water and the released heat is used to evaporate the feedwater.

23

Case Studies

In this chapter we present two case studies. The first is a complete water audit report undertaken by a student in our teaching unit at Murdoch University. The purposes of including this case study are as follows. It provides an illustration of the ordering and structuring of a water audit report as outlined in Chapter 5. Also it provides an illustration of what imaginative thinking can achieve when developing a water management strategy for a very simple audit indeed. We ourselves are not necessarily recommending the strategies devised, but we strongly affirm the process that has led to them and the thorough way the options are recorded. The second case study covers the water treatment plant and the steam cycle of a substantial manufacturer, who wishes anonymity for commercial reasons. We have massaged the case study to conceal identifying features. It is not possible to present the whole report both because of its size (50 pages) and because presenting such detail is inconsistent with confidentiality. We have presented the essentials, which enable us to observe typical savings of both water and money, creative aspects to the water management strategy and the relationship between water auditing, environmental auditing and environmental management systems. In doing this we illustrate material from Chapters 1, 2 and 6 as well as the water auditing process itself.

23.1 A CASE STUDY ON DOMESTIC UNITS

We present this case study as a fully written up report, including the title page. (see below.)

WATER AUDIT OF UNITS '2' AND '3', CITY OF MELVILLE,

PERTH, WESTERN AUSTRALIA.

Conducted by
Teneal Davidson.

From (date suppressed) until (date suppressed)

For

Client: Jeff Sturman, and

Auditees: names suppressed

Date of report: (suppressed)

Participants

Eight individuals and three companies; names suppressed

EXECUTIVE SUMMARY

Four years after the proposed water management strategy has been implemented all the items required for saving water in Unit 3 will have been "paid for" by the water cost savings incurred. Subsequently the annual water-savings are expected to be greater than 142kL which, at a water rate of \$0.379/kL, amounts to \$53.82 saved per annum.

The proposed water-savings for Unit 2 will have paid for the water saving items within 2 years. After this time the annual water-savings are expected to be greater than 215kL or (at a water rate of \$0.613/kL) \$132.85. These savings could be applied to any Raw Options, which appeal to the homeowner, or to other unspecified areas.

The capital requirements for implementing the water management strategy amount to \$200 for Unit 3 and \$280 for Unit 2 - not including the cost of plants needed for the windbreak (see Table 1). The cost involves the minimum prices of items such as rain-gauges and electric razors. The expense and capabilities of the expense items is the prerogative of the homeowner.

If water reuse options are implemented the environmental benefits for the household can be substantial. Not only are they saving water and money for the household but when used on the garden and lawns, the reused water helps to recharge local groundwater aquifers.

INTRODUCTION

The domain for both Units 2 and 3 consists of a 2 storey townhouse on a $190m^2$ ground area (including the gardens and verandah) with a double garage beneath. Both units contain 2 toilets, 2 hand basins, a laundry sink, a kitchen sink, a dish washer, a washing machine, a spa, a bath, 2 showers, 2 external faucets, and an irrigation system with 8 full spray sprinklers, 15 half spray sprinklers, and 4 drippers (for hanging plants).

Unit 2 contains 5 people, and Unit 3 houses 3 people. The domain currently has few water-saving devices and the inhabitants apply few water-saving practices. Changes in water resource availability are unlikely to change in the near future, but if household water-use is not minimised in Perth, there may be regulations applied and water use restrictions.

A current incentive to use less water is the reduced water-rate for the first 150kL of water used by a household in a 6 monthly period (\$0.379/kL). After this amount the rate is increased to \$0.613/kL, in order to discourage users from going over the 150kL limit. Unit 3 uses less than 150kL in 6 months so they are never charged more than \$0.379/kL, whereas Unit 2 uses approximately 81kL over and so is charged \$0.613/kL after the 150kL each 6 months.

The unit complex houses 5 individual townhouses on what was originally a single house site. The previous domain would have used much less water within the building, and more on the lawns and gardens. So the change in water sinks has probably been dramatic. Formerly, the majority of the water used in the domain would have been used on the garden and that which didn't evaporate eventually ended at an underground aquifer. Now the majority of the water enters the sewer as it has been used for household purposes.

As the domain is domestic housing the financial savings tend to be limited. The more costly water-saving options are not cost effective and have very long payback periods. Thus the main onus is on implementing behavioural changes, which will reduce the water consumption of both households.

WATER SOURCES

The only current water source is scheme water, which does not have a limit on the quantity used each year. Even so scheme water costs $0.379 for the first 150 kilolitres (kL) used in a six month period, and $0.613 for each kilolitre used after in that six month period. These charges are used by the Water Corporation as a regulatory control, so that if you only use 150 kL per six month period then each kilolitre is $0.368 cheaper than each kilolitre of water in excess of 150 kL.

If the household water usage can be reduced in unit 2 by 165kL per annum then the water bill cost would become $113.70, a reduction from the current water costs of $101.15 (City of Melville Rate Notice, 2001).

Potential water sources include bore water, rainwater and "grey water". Bore water is only available if there is access to an aquifer on the domain site; this is quite likely from local groundwater hydrological knowledge and can be proved by a professional driller. The cost of a bore (including the detection, drilling and pumps) is between $3000 and $6000, depending on the depth needed and the power of the pump required. Charges are not levied on the use of the groundwater, so running costs are confined to the electricity required to run the pump and maintenance.

"Grey water" is water collected from washing clothes, from showers and hand basins (Addkison and Sellick, 1983) (In Perth dishwashing water is classified as black water because of likely meat content.) Greywater can be used to flush toilets and to water gardens and lawns if applied sub-surface. If it is being used on the garden or lawn then biodegradable products should be used in the washing of clothes.

Rainwater that falls on the roof and on paths can be collected from down-pipes and drainage pipes beside paved areas, for storage and later use (Addkison and Sellick, 1983). Rainwater from roofs can be used for drinking and cooking, with some minimal filtering. Water collected and stored from paved areas can supplement normal garden watering or irrigation. Other options include flushing toilets and washing clothes and dishes (depending on quality).

WATER SINKS

Current sinks are sewers (which receive 91% and 89% of used scheme water from units 2 and 3 respectively), stormwater drains (which receive approximately 2% of used scheme water from both units), and the ground (which receives 7% and 9% of used scheme water for units 2 and 3 respectively).

The sewers receive far too large a percentage of scheme water from both units, which is an issue that needs to be addressed. Water from dishwashing, clothes washing, showering, hand-basins and washing or cooking vegetables can be used on the garden thus help recharge the groundwater supply (Addkison and Sellick, 1983; Randall, 1977; and Walsh, 1993). This practice could result in around 60% to 70% of the total scheme water used recharging the local aquifers. If car washing was carried out on the lawn at the front of the complex (although it is a little awkward), then around 2% of scheme water would be directed to recharging the groundwater, rather than to stormwater drains.

Detergents and other contaminants (such as bacteriological and cooking oils/fats) within "grey water" are removed by filtration through sand and limestone (Melville is within the Spearwood Dune System – sand overlying limestone - and lies between the Karrakatta and Cottesloe soil association, evident from the leached yellow sands in the area; Lantzke and Venkitachalam, 1997) before it reaches the aquifer (Porter and Skinner, 1995).

RECYCLED AND RE-USED WATER

"Grey water" (See above.) This water source costs nothing.

FLOW MEASUREMENT

Scheme water supply was measured using the water service provider's cumulative flow meter over the period of the audit. The water use in the audit period was compared with the audit period's appropriate proportion of the annual water bill water quantities, to ensure that the values in the audit period were representative. The two matched within 10% and this was regarded as adequate. Water usage within the units was determined using 'bucket and stop watch' unless otherwise noted. Toilet and shower use frequency were obtained by a pen and 'score' sheet located in the toilets. Shower duration was entered on a score sheet, the occupants using a timer located at the showers. Dish washer water usage per load was obtained from manufacturer's data sheet. A score sheet was also used to determine frequency of usage. The results are summarised on the water flow diagrams of Figures 1 and 2.

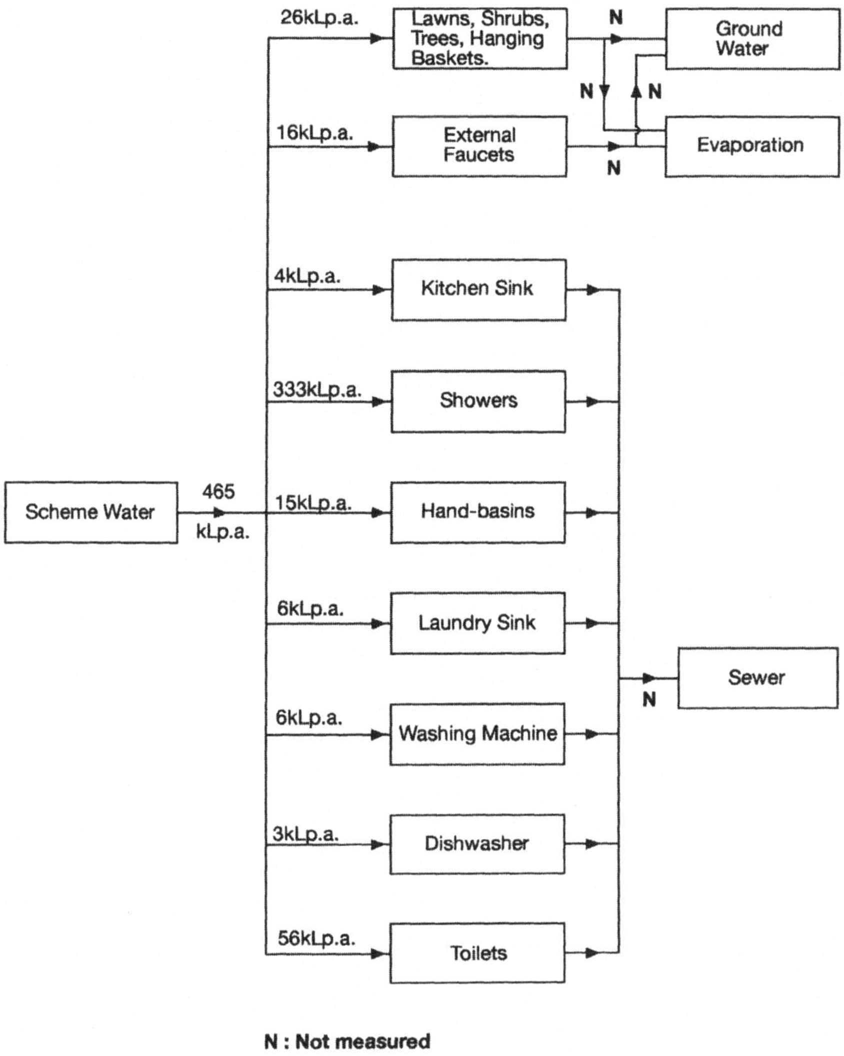

Figure 1 Water flow diagram for unit 2.

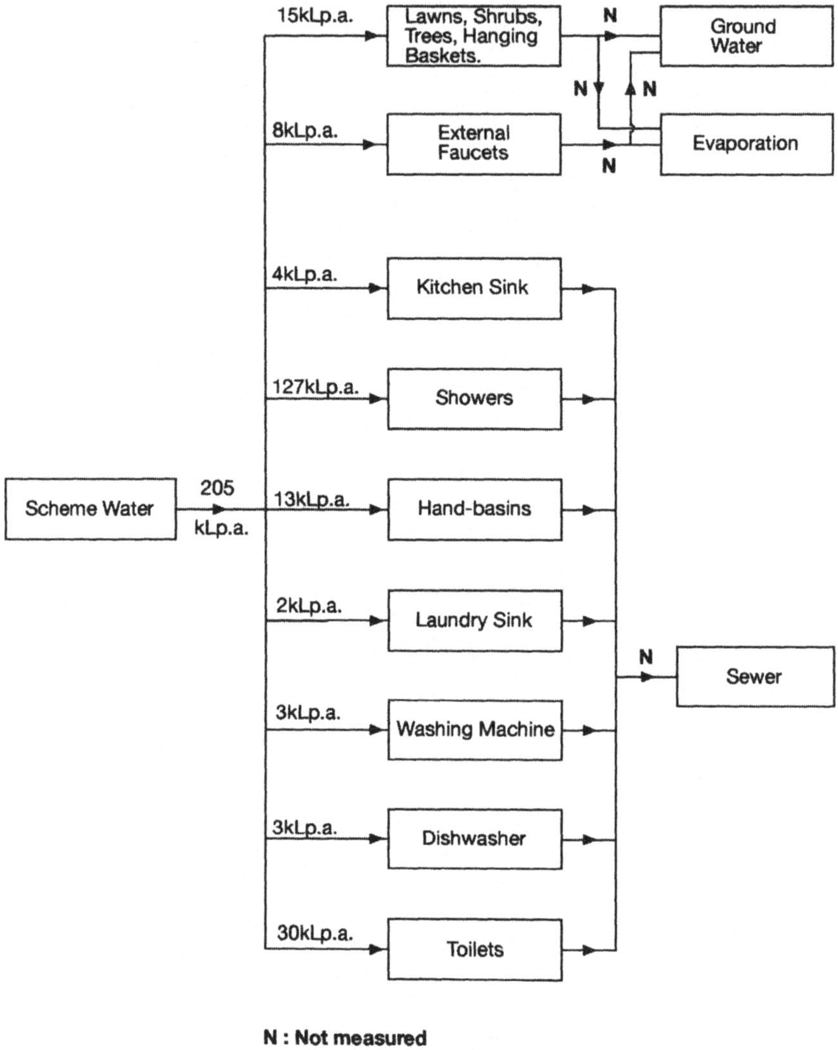

Figure 2 Water flow diagram for unit 3.

CLOSURE

For unit 2 the total scheme water input was 465 kLpa, while the total of individual water usages was 439 kLpa. This results in closure to within $(465-439)\times100/465=6\%$. This is less than 10% and is satisfactory.

For unit 3 the total scheme water input was 205 kLpa, while the total of individual water usages was 205 kLpa. This results in closure of 0%. This is unusual and fortuitous.

RAW OPTIONS FOR IMPROVED WATER USE EFFICIENCY

Behavioural Changes to Improve Water Use Efficiency

- Use the washing machine only when there is a full load.
- Wash cars on the lawn at the front of the complex.
- Don't wash fruit or vegetables under a running tap. Fill a bowl, rinse vegetables/fruit and then pour water onto garden.
- Don't leave tap running when shaving, brushing teeth, or soaping-up soiled hands. Fill a basin a little and use this water. This practice can save 43 litres or more per person, per day. (The Metropolitan Water Board and Public Works Department, 1979)
- Change showering practices. That is, wet down, turn off shower, soap up, rinse off. This practice uses only 18 to 20 litres of water.
- Only use the dishwasher on "Economy" and when it is fully loaded.
- Don't rinse dishes before putting them into the dishwasher. If food is stuck on then soak stacks of dishes at a time.
- When leaving on holiday turn off the main water supply. That way if something were to spring a leak it wouldn't go unnoticed.
- Landscape gardens with native plants, as they generally require less water.
- Add loam to sandy soils as it absorbs and retains water for your plants.
- Remove weeds so they don't rob plants of water.
- Turn off the automatic irrigation system if ti has been raining, or install a soil moisture sensor.
- Don't use sprinklers on a windy day. Drift will redirect substantial amounts of water, and plants won't receive the benefits.
- Water before 8am.
- Don't use the hose as a broom.

- When finished with the hose turn it off at the tap to prevent leakage. (Harrison, 1997)
- Cook vegetables in steamers rather than boiling.
- If boiling vegetables, save the water to use as vegetable stock. The more times it's used the stronger the flavour.
- When cooking rice or pasta let the water cool then pour on outdoor plants.
- Use wet wipes instead of washing hands.
- Fix leaks as soon as possible, in the meantime shut off the water supply when not in use.
- Mulching of gardens:
    - decreases the amount of water lost through evaporation
    - keeps the soil cooler
    - keeps soil porous, allowing good aeration and water to reach the plants' roots
    - adds organic matter to the soil
    - protects shallow-rooted plants from excessive drying
    - helps to control weed development
    - apply mulch 5 – 10 cm thick
- To conserve water cut the lawn high – 8 to 14 mm.
- Group plants according to their water needs; (hydrozoning).
- Irrigate efficiently, that is don't let sprinklers water paths, or space sprinklers too closely (they should be one radius of the sprinkler throw apart; UM, 2002, p135).
- Check the soil to see if it is damp before you water.
- Aerating the lawn will improve water penetration too.
- Hand-watering of potted plants is the most efficient, as the amount can be easily judged. Do not keep applying water to the top of the pot while it is draining out of the bottom. Alternatively, pots of manageable size can be plunged in a container of water and drained before being returned to their location.

Structural Changes to Improve Water-use Efficiency.

- Change current "rose" showerheads to the reduced-flow type. Cost: $10; saves 9L/min. (M.W.B.P.W.D., 1979; Beri Distribution [pers.comm.], January 2002).
- Connection of strata units to groundwater-bore system for use in irrigation rather than scheme water. Cost: $3000 - $6000; can save up to $21 on water bills per unit, per year. (M.W.B.P.W.D., 1979; Glen B&L Pumps [pers.comm.], January 2002).

- Explore the possibilities of drip irrigation and soil soakers. Easy to install, low cost (pay for the irrigation pipes), and reduced evaporation.
- Avoid fine sprinkler sprays (current type) especially for open areas such as lawns.
- Buy an electric razor. It saves water and gas to heat the water, very little electricity used. Cost: from $80 (Kmart, January 2002). (Harrison, 1977).
- Collection and storage of rainwater. Prices for tanks depend on size. Pipe costs depend on lengths required.
- Soil moisture sensor from garden supply stores determine the moisture levels in the soil. Cost: $80 - $90. (Addkison and Sellick, 1983).
- Lawn irrigation should be on a separate irrigation zone to gardens, as each have different watering requirements.
- Drip irrigation is the most efficient watering system. (Walsh, 1993).
- A windbreak across the open-railed fence would reduce the amount of sprinkler-irrigation drift. Use of low-maintenance, drought-tolerant climbers is suggested.

EVALUATION OF RAW OPTIONS

Behavioural changes

Behavioural changes are the easiest to implement, but must be chosen to fit-in with the house-occupants lifestyles. They can also be less durable than structural changes. Table 1 provides an evaluation of the most easily achievable options that have no initial cost requirements.

Options Requiring Finance to Implement

Some of these options, namely the soil moisture sensor and electric razor options (Table 2) have a considerable payback period (the period of time taken for the savings in operating costs to payback the initial capital cost). However, this does not mean that they are less important. The savings achieved from the "non-cost" changes in water use can be used to pay for the expensive changes such as mulching, planting of wind-break, showerheads, soil moisture sensor, and electric razors.

Table 1(a): An evaluation of the most easily achievable options in relation to water, environmental, human and financial issues.

Easily Implemented Options	Water Issues	Environmental Issues	Human Issues	Financial Issues
1. Use washing machine only when there is a full load.	Less water used more efficiently. Can be used as "grey" water.	Less electricity used – fewer emissions. Water reused for gardens - recharge groundwater supply.	May take time for water users to readjust practices.	Water and electricity savings. Water savings up to 0.8kL p.a. (approx. $0.74 p.a.)
2. Wash cars on the lawn at the front of the complex.	Quality of water is middle to low, but impurities are filtered by movement through sand and limestone.	Water won't pollute lakes and waterways via stormwater drains. Water recharges the local aquifer.	Requires a little more effort (transport water via buckets or hoses).	Financial savings if "grey" water is used, between 4 and 8kL of water can be saved. Financial saving of $3.70 to $7.40.
3. Change showering practices. Wet down, turn off shower, soap up, rinse off.	Uses only 18 to 20 litres of water per shower. Water can be collected for use on garden.	Less water entering the sewers. If collected can help to recharge the ground supplies.	May take time to adjust to and implement. Shorter shower time, more time in the mornings. Might be unacceptable in cold weather	Up to 20 L/shower saved per day. U2: 20L x 8(showers)= 160L/day x 365 / 1000 =58.4kL/year saved ($35.80). U3: 20L x 5(showers)= 100L/day x 365 / 1000 =36.5kL/year saved ($13.83).
4. Only use dishwasher on "Economy" cycle and when it is fully loaded.	The most dishes cleaned with same amount of water. Eco uses least water to remove food and sanitise dishes.	Allows cooking fats and oils, detergents and biological matter into sewer system. Can be used on garden.	Stack dishes in dishwasher to be washed later rather than filling a sink for just a few.	Eco saves between 4 and 5 L/ wash. Dishwasher can save around 30L/week (if sink is filled once a day) $1.50p.a.
5. When finished with the hose turn it off at the tap.	Prevents wastage of scheme water.	Less water taken from the scheme-dam.	Little difficulty.	Variable. Depends on the size of the leak and the rate.
6. Displacement bottles (1/3 sand/pebbles, 2/3 water) in toilet water-tanks.	Reduces the amount used in flushing. Doesn't affect functioning.	Reduces amount of water entering sewers, and scheme water used.	Takes little effort and can save a lot of water.	Can save up to 4L/flush. 17.3kL saved/year ($10.60) for Unit2, and 9.4kL saved/year ($3.60) for Unit 3.

Table 1(b): An evaluation of the most easily achievable options in relation to water, environmental, human and financial issues.

Easily Implemented Options	Water Issues	Environmental Issues	Human Issues	Financial Issues
7. Mulching of gardens.	Reduces evaporation; allows water to reach the plants' roots; less water needed to keep plants healthy.	Keeps the soil cooler; adds organic matter, improving soil; protects shallow-rooted plants from excessive drying; helps to control weeds.	Reapplication usually required annually, depending on rate of breakdown of material used; reduces irrigation requirements.	Prices per 20kg bag of mulch range from \$20 to \$60 depending on the type of mulch used. Water savings are unable to be determined.
8. Check soil moisture before watering.	Watering of the garden should be regulated to suit the plant's needs and the water-holding capacity of the soil. Not needed after rain.	Reduces water-logging of the soils; helps to reduce over-use and wastage of scheme water.	Test for moisture 2cm into the soil, below mulch. A soil moisture sensor measures soil moisture, and can be connected to the irrigation system to determine watering requirements.	Manual moisture testing costs nothing. Soil moisture sensors can cost between \$80 and \$90 (Hugall and Hoile, pers. comm. January 2002)
9. Install a windbreak across the open-railed fence.	Reduces evaporation and loss of water through wind-drift. If low maintenance, drought tolerant climbers used, they're likely to use much less water than that lost to wind drift.	Helps to reduce over-use and wastage of scheme water; can protect less hardy plants from sun and wind damage; can reduce the temperature of the garden's microclimate; aesthetically pleasing.	Plants may need to be tied in place in the first few weeks, and given special attention whilst establishing (eg, extra water and fertiliser). (walsh, 1995)	Initial cost of the climbing plant chosen (prices vary according to species). Water redirected from wind-drift to the garden is difficult to determine, but may be detected through the use of tensiometer readings once the windbreak is established.

Table 2: Evaluation of options requiring an initial expense outlay in relation to water, environmental, human and financial issues.

Options Requiring Financial Assistance	Water Issues	Environmental Issues	Human Issues	Financial Issues
1. Conversion to reduced-flow showerheads.	Flow rate reduced by 9L per minute. Water wetting quality and shower intensity are not compromised.	Helps to reduce over-use of scarce scheme water.	Hole-size can be manually adjusted to suit preference. Easy to install.	Initial cost of approximately $10, or $20 for 2 (Beri Distributors, pers. comm. January 2002) – prices vary between stores and styles. Would save approximately 105.1kL p.a. ($64.44) for Unit 2, and 65.7kLp.a. ($24.90) for Unit 3.
2. Installation of a soil moisture sensor.	Amount of high to middle quality water applied to the garden and lawns is according to plant need.	Minimises unnecessary use of water when soil is already moist	Eliminates human judgment about when to water..	Cost is around $80 to $90 (the more high-tech the greater the expense). Can save more than 8kL per year just on rainy days; 22kLp.a. for Unit 2 ($13.50 at $0.613/kL) and Unit 3 ($8.34 at $0.379/kL).
3. Use of an electric razor instead of a manual razor, which uses water.	A full sink for shaving uses around 10L of water per shave. An electric razor performs a dry shave, thus not needing shaving foam or water.	Uses very little electricity and no scheme water. Does not deliver shaving foam into the sewer systems.	Saves time and money – on replacement razors and shaving foam. Less likely to cut oneself.	Initial cost is $80 minimum (Kmart, pers. comm. January 2002). Can save at least 2.4kLp.a. ($1.50) for Unit 2 and 1.2kLp.a. ($0.47) for Unit 3.

Total Savings Per Annum (On "Non-Cost" Options)

Table 3: Individual water savings for the non-cost options, incorporating mostly behavioural changes, to be implemented.

Option	Unit 2	Unit 3
1.	$0.50	$0.30
2.	$4.90	$1.52
3.	$35.80	$13.83
4.	$1.60	$0.91
6.	$10.60	$3.60
	Total Saved: $53.40p.a.	**Total Saved: $20.16p.a.**
	Current annual water cost $217.00	Current annual water cost $76.96
	Cost after the above options are implemented: $163.60	Cost after the above options are implemented: $56.80

Total Outlay for Expense Options and Payback Periods

Table 4: Water savings and initial outlay for expense options for implementation.

Option	Unit 2 capital cost	Unit 2 Saves in water use	Unit 3 capital cost	Unit 3 saves in water use
Mulch	$20		$20	
Showerheads	$20	$64.45	$20	$25
Rain-gauges	$90	$13.50	$90	$8.34
Electric Razor	$160	$1.50	$80	$0.47
Totals	$280	$78.57	$200	$33.27
Payback period		3.6yrs		6 yrs

Reuse Options and Water Savings

(1) Using "grey-water" from processes such as washing vegetables, dishes and cooking; showering; hand-washing and shaving; Laundry, including clothes soaking and washing machine; and dishwasher. This water can be used for flushing toilets, watering the lawn and garden, and washing cars. Reusing water in this way can save up to 66.7kL per annum for Unit 2 and 35.6kL per annum for Unit 3 if the water is used in the same way as proposed in the evaluated options and proposed water flow diagrams.

(2) Rainwater collection can be costly to implement and requires a certain amount of space. Below ground tanks are an option that doesn't require a lot of above-ground space. But for both Units 2 and 3 the option is not plausible as both have very little room and soil depth in the backyards is too shallow to support a rainwater collection device. An alternative may involve pipes carrying collected water to the garages beneath the two units where a rainwater storage container could be housed. This would not be difficult as the down-pipes for both units terminate nearby. If this could be redirected to a storage container then there would be an alternative source of high quality water to be used for various applications.

WATER MANAGEMENT STRATEGY

Both Units 2 and 3

Behavioural changes to be implemented by the end of the month (beginning 1st March 2002):

- From Table 1, options 1,2,3,4,5,6,8.

Minimal expense options can be implemented within the month of March. From Tables 1 and 2:

- Changing the current showerheads to reduced-flow ($10)
- Mulch the garden beds ($20 for 20kg)

Expense options to be implemented by the end of 2002:

- Planting and training climbing plants (varying cost) across the open-railed fence to form a wind break.
- Plants suggested – Bougainvillea (Bougainvillea glabra), Happy Wanderer – Native (Hardenbergia violacea), Japanese Honeysuckle (Lonicera japonica), Wonga Wonga Vine – Native (Pandora pandorana), Australian Bluebell Creeper – Native (Sollya heterophylla). (Walsh, 1995)

Unit 2 After 1st Year

Water savings should amount to at least 192kL ($117.85) this money can be used to purchase a rain-gauge to connect with the automatic irrigation system, saving at least 20kLp.a. ($12.62).

Unit 2 After 2nd Year

Water savings should amount to at least 212kL ($130). This can contribute to the purchase of 2 electric razors (minimum cost $160). These will save around 2.5kLp.a. ($1.50) as well as saving on time razor and shaving-foam costs.

Unit 3 After 2nd Year

Water savings should amount to at least 238kL ($90.32). This money can substantially purchase a soil moisture sensor for connection to the automatic irrigation system, saving at least 20kLp.a. ($7.80).

Unit 3 After 4th Year

Water savings should amount to at least 278kL ($105.36). This money can purchase an electric razor and still save $25.36 (if minimum purchase cost remains $80 after 4 years).

NB: These estimations do not take into account any changes in either water rates or costs of items.

REFERENCES

Addkison, R. and Sellick, D. 1983. *Running Dry: how to conserve water indoors and out.* Stein and Day, New York.

City of Melville Rate Notices, P.O. Box 9000, Applecross, W.A., 6953.

Harrison, R. 1977. *How to cut your water use in half and still stay sane and sanitary.* Communication Press, San Francisco.

Lantzke, R. and Venkitchalam, H. 1997. *Introduction to Environmental Science: Unit Readings.* Murdoch University, Murdoch, Perth.

Pilatowicz, G. 1995. *Eco-interiors: a guide to environmentally conscious interior design.* Wiley, New York.

Porter, S. and Skinner, B. 1995. *The Dynamic Earth: An Introduction to Physical Geology.* John Wiley and Sons, Inc., Canada. pp.280 – 306.

Principles of Water Conservation and Water Auditing: external unit guide. 2002. Murdoch University, Perth, W.A.

Promoting the efficient use of water resources through management of demand. 1986. Western Australian Water Resources Council, Perth.

The Metropolitan Water Board and Public Works Department. 1979. *Resource 1: an exposition on water.* Public Works Department of Western Australia, Perth.

Walsh, K. 1995. *Water-Saving Gardening: water-wise plants and practices in Australia.* Reed Books, New South Wales.

23.2 A CASE STUDY ON THE WATER TREATMENT AND STEAM CYCLE OF A MANUFACTURING PLANT

The subject of our case study is the water treatment plant and steam cycle of a substantial manufacturing plant. Historically the plant's management have had a positive attitude towards environmental auditing and water conservation. We quote here some pertinent extracts from their environment policy[1].

"To ensure environmentally responsible behaviour is embraced as an integral part of our operations, we specifically undertake to:

- conduct operations in compliance with relevant local environmental legislation, regulations and licenses,
- prevent pollution and achieve continual improvement of environmental performance through economically viable best practice,
- regularly set and review environmental ... objectives and targets
- educate our people and contractors, ensuring the requirement for environmental responsibility is integrated into work practices ..."

Though presented in a different form, the policy quoted contains all of the elements of the environmental management system model discussed in Chapters 2, 6 and ISO 14004:1996 and ISO 14001:1996, with the exception of explicit reference to measurement and evaluation.

From their water resource management policy we quote:

" (Name of company) is committed to responsible water management to minimise the total cost and overall impact of water usage at (location of plant). Water usage at (location) will be minimised to:

- Improve cost effectiveness on a total life cycle approach,
- Reduce overall environmental impact.
- "Water prices for (plant name) will be minimised by a pro-active program of contract "negotiation and continued assessment of alternative water sources, providers and opportunities.

Key factors of the Water Policy will be:

- communication of the policy to all levels on site ...
- accountability for water usage management coordinated by a 'Water Coordinator' reporting to the manager
- detailed monitoring of water use and review against targets
- regular communication on water to all employees
- promotion of the benefits of water efficiency and training in ways to achieve it
- implementation of a water reduction plan to be reviewed bi-monthly
- investment in projects and cost effective opportunities for water efficiency."

---

[1] We do not cite the documents, as the company does not want to be identified in the public arena.

By contrast, the key factors of the water policy specifically include monitoring and review against targets. This is consistent with an environmental audit (ISO 14010:1996), where measurements are compared with criteria. The water policy contains within it all the key elements of water auditing, though not systematized to reflect the contents of Figure 2.3. We note that the policy specifically commits to water conservation in the item 'implementation of a water reduction ...' and is implied in other items. Thus the company culture was favourable to the notion of a formal water audit. We will now describe the water audit before interpreting its process and outcomes as an exemplar of a type of environmental management system. The audit itself was undertaken as a final year undergraduate project (Levett & Weatherburn, 2001)

The site of the plant is on the sandy coastal plain of Perth, Western Australia, with a Mediterranean climate. The Water Corporation Water supplies water from dams in the hinterland hills, where the rock substrate is granite. The water is reticulated, treated water of high quality. On the coastal plain there are also confined and unconfined sub-surface aquifers from which bore water can be drawn. By contrast with the scheme water costing about AU$0.70 per kilolitre, bore water can be accessed in large quantities for the cost of installing and maintaining a bore and the cost of powering pumps. Water was extensively recycled or reused prior to the audit. Both scheme water and bore water are treated prior to use both in boilers to raise steam and to use in the manufacturing process. The audit was confined to the water treatment plant and to the boiler and steam circuits. From this arena water was discharged to the manufacturing plant, to irrigation, to evaporation and to the product.

A simplified water flow diagram is presented in Figure 1, showing input and output water flow rates that were either directly measured or inferred from direct measurements. Extensive measurement was taken inside the large bare box, but this is omitted here for the sake of brevity and simplicity. There are recycled or reused water loops within the bare box too. A closure calculation was attempted using the inputs and outputs and it was obtained to within 9.5%, an acceptable result falling within the industry norm of 10%.

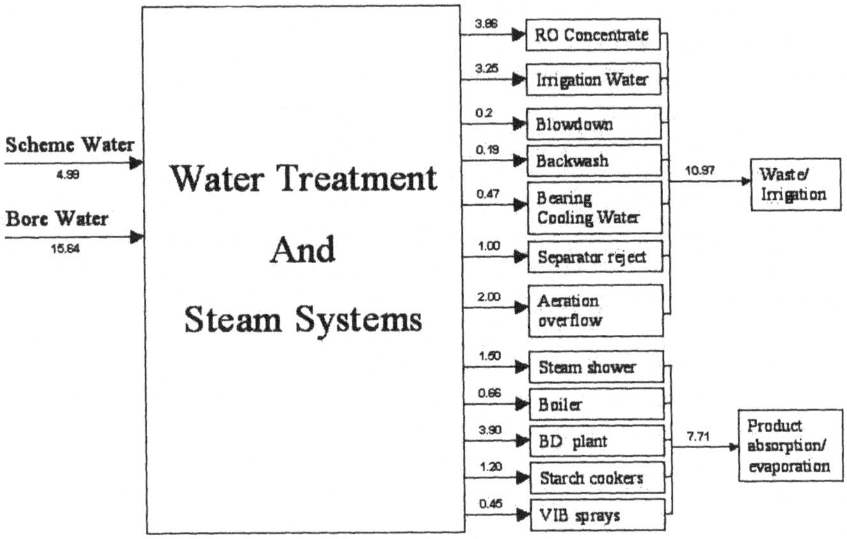

Figure 1 Simplified water flow diagram for the water treatment and steam systems. Units of flow are m^3/hr (kL/hr). (Source: Levett and Weatherburn, 2001)

With the assurance of a reasonable understanding of processes and water flows, a range of options for reducing water use, recycling or reusing water or substituting new sources were listed as a result of a brain-storming process. Financial, human and environmental factors were considered in evaluating these options. Those that had merit by all criteria were more carefully evaluated financially. A short list of these options was then developed into a water management strategy. The results are summarised below in Table 2. Given that the management of the auditee company implements the recommendations of the water management strategy as summarised in Table 2, the water audit can be seen to be an environmental management system. ISO 14001:1996 and ISO 14004:1996 highlight the cyclical repetitive nature of water management strategies. The final recommendations, if implemented by the plant management, complete the factors necessary. Finally we note a total financial benefit to the company of approximately \$9500/yr and total water saving or beneficial use of earlier wasted water of 79.9ML/yr. Though the financial saving is not large, it relates to only a small portion of the plant.

In water auditing we have the coincidence of an intellectual mechanism of Abu-Zeid's (1998) vision of water and the world, of environmental auditing and of environmental management systems. The practice of water auditing is thus a powerful tool, which is diminished by the common reduction of water auditing to a mere quantification of water flows on a flow diagram.

Table 2 Summary of Water management strategy recommended actions

Recommendation	Capital cost	Payback period	Comments
Feedwater pump bearings – recycle cooling water	$2,300	Less than one year	Pumping details need refinement
Plant Blue-gums on land currently irrigated but non-productive	$10,500/10yrs	10 yrs	Profit of $14 500 after 20 yrs. Other benefits: buffer zone, future carbon credits and aesthetics.
Recover steam separator water	$7000	5 months	
Store boiler water from maintenance shutdowns			This is an uncosted option, saving very high quality treated water of small volume.
Conduct further water audits in other parts of the plant. Repeat all audits periodically. Update existing measurement systems.			This recommendation is based on the success of this audit. Periodic audits are necessary to continually refine water management strategies.

References

Abu-Zeid M. A. (1998) Water and Sustainable development: the vision for world water, life and the environment. *Water Policy*, **1**, 9 – 19.

AS/NZ ISO 14001:1996 *Environmental management systems – specification with guidance for use.* Standards Australia and Standards New Zealand.

AS/NZ ISO 14004:1996 *Environmental management systems – General guidelines on principles, systems and supporting techniques.* Standards Australia and Standards New Zealand.

AS/NZS ISO 14010:1996 *Guidelines for environmental auditing – General principles.* Standards Australia and Standards New Zealand.

Levett, J. & Weatherburn, A. 2001. *N421 Environmental Science project.* School of Environmental Science, Murdoch University, Perth, Western Australia.

Index

IWA Publishing's authorised EU representative for General Product Safety Regulations is Diane D'Arras, 15 rue Duret, 75116 Paris, France, e-mail: safety@iwap.co.uk.

Printed and bound by CPI Group (UK) Ltd, Croydon, CR0 4YY
12/05/2026
02108867-0003